The Sustainable Use of Concrete

The Sustainable Use of Concrete

Koji Sakai

Takafumi Noguchi

CRC Press
Taylor & Francis Group
Boca Raton London New York

CRC Press is an imprint of the
Taylor & Francis Group, an **informa** business

CRC Press
Taylor & Francis Group
6000 Broken Sound Parkway NW, Suite 300
Boca Raton, FL 33487-2742

First issued in paperback 2017

Version Date: 20120703

ISBN 13: 978-1-138-07588-7 (pbk)
ISBN 13: 978-0-415-66720-3 (hbk)

Library of Congress Cataloging-in-Publication Data

Sakai, K. (Koji)
   The sustainable use of concrete / Koji Sakai, Takafumi Noguchi.
      p. cm.
   Includes bibliographical references and index.
   ISBN 978-0-415-66720-3 (hardback)
   1. Concrete. 2. Concrete construction. 3. Sustainable construction. 4.
Concrete--Environmental aspects. I. Noguchi, Takafumi. II. Title.

TA439.S24 2012
624.1'834--dc23   2012014558

Visit the Taylor & Francis Web site at
http://www.taylorandfrancis.com

and the CRC Press Web site at
http://www.crcpress.com

Contents

Preface

The construction industry is very conservative. This can be seen as deriving from the special nature of its work, which is creating the social and economic infrastructures required by each particular age in a "safe" way. Architecture is to some extent ahead of its time in the design of buildings, but also reflects the inclinations of clients. In other words, the basic activity of the construction industry has been to reliably translate social needs into material form. Naturally, with the growing sophistication of requirements, construction technology has developed and many breakthroughs have been achieved to make the impossible possible, but this process has also been marked by many failures. The construction industry can be said to have built its technology systems through a process of "experience engineering."

The construction industry exhibits a high degree of locality. Structures have generally been built by local people using local materials. Globalization has promoted internationalization in the construction industry as elsewhere, but the basics of construction systems have remained unchanged. What makes this possible is the wide use of concrete as a construction material. Its primary component materials are aggregate, cement, and water, with aggregate constituting approximately 70% of the total volume. The Earth's crust is composed of rocks that are the raw materials for aggregate. Water, when seawater is included, is the most commonly available globally circulating substance on Earth. In the case of cement, the raw materials are limestone and clay, both of which are also available in abundance. Limestone consists of carbon dioxide—present in huge quantities in the atmosphere of the young Earth—fixed biologically and chemically as calcium carbonate that is also present in huge quantities. Clay is formed by the chemical reaction of rock and water. Concrete is thus made from the most abundant substances on Earth, and this is one of the main reasons why its production has been able to expand to respond to growing construction demand. Concrete is now the second most consumed substance on Earth after water. To state the case in extreme terms, we might say that contemporary society could not exist without concrete, and it is no exaggeration to say that the development of a nation is directly correlated to its consumption of concrete.

It is not clear how the construction industry reacted when the Rio Declaration, intended to satisfy both environmental conservation and development, was adopted at the Earth Summit held in Brazil in 1992, but it seems unlikely that the expected result would be stagnation of the construction industry. This is because the declaration's general description of environmental conservation was abstract, and in particular, focused on the classic problems of environmental pollution.

When the Kyoto Protocol was adopted in 1997, the reactions of some organizations in the construction field were relatively swift. As a result, assessment and certification systems for building sustainability are now in place, but many of them concentrate mainly on building operation concerns while little consideration is given to the environmental burdens caused by construction materials themselves. This has meant that action to reduce environmental burden in the area of materials has lagged. Meanwhile, as energy efficiency in the field of building operation has begun to improve dramatically, the proportion of environmental burden deriving from materials is beginning to show an upward trend in relative terms. Particularly since the consumption of concrete is expected to continue to rise markedly in the future, the concrete industry has begun to see that it needs to take some kind of action. In countries around the world, concrete-related institutions and organizations have begun work aimed to assess current conditions and develop technology. However, we are still at the stage of trying to find a solution through a process of trial and error to determine exactly what we should do and how.

The authors have been engaged in leading a range of activities relating to the environment and sustainability in Japan through the Japan Society of Civil Engineers, the Architectural Institute of Japan, and the Japan Concrete Institute and at the international level through Commission 3 (Environmental Aspects of Design and Construction) of the International Federation for Structural Concrete (fib), TC71/SC8 (Environmental Management for Concrete and Concrete Structures) of the ISO, the ACI Concrete Sustainability Forum, and the ACF Sustainability Forum. The authors pride themselves on having been consistently at the forefront in issues concerning the environment and sustainability in the field of concrete, and consider it their duty to society to analyze the current situation regarding sustainability in the concrete industry and map out prospects for the future. This background led them to respond to a request from Tony Moore of Spon Press to write this book.

The book devotes a considerable number of pages to the background and thinking behind sustainability. This decision was based on the fact that many concrete engineers and researchers tend to be concerned only with the technology and spend little time trying to understand the essence of the problem. It was also out of the authors' desire to share with the reader their understanding that the concept of sustainability amounts to the equivalent of a third revolution in which we will need to fundamentally rethink the results of mankind's past achievements during the Agricultural and Industrial Revolutions.

The content of this book does no more than mark the start of efforts to realize sustainability in the concrete and construction industry. If it can help the succeeding generations during the course of the future development of such sustainability, it will have more than fulfilled the authors' hopes.

Finally, the authors express their enormous gratitude to Tony Moore, who gave them the opportunity to write this book, and to the editors Siobhán Poole and Prudence Board who provided valuable advice on the manuscript.

They would like to thank Ms. Miki Ido, Yoshiko Ogawa, and Miss Dayoung Oh for their dedicated work as assistants to both authors.

Special gratitude goes to Dr. Liv Haselbach, Washington State University; Dr. Hiroshi Kasai, Kajima Corporation; Prof. Masashi Koyanagawa, Tokyo University of Agriculture and Technology; Dr. Yasuhiro Kuroda, Shimizu Corporation; Dr. Kohsuke Yokozeki, Kajima Corporation; Mr. Hiroyuki Musha, Taisei Corporation; Dr. Mitsuhiro Ishii, Shikoku Research Institute; and Mr. James Prowse, Gifford (now part of Ramboll Group) who provided valuable information for case studies. The authors also express their heartfelt thanks to all those with whom they have worked in the past.

Koji Sakai
Kagawa University

Takafumi Noguchi
The University of Tokyo

1

Introduction

The origins of concrete go back to the Greco-Roman era (Malinowski 1979, Davidovits 1987), but some research suggests that a prototype form existed in the Neolithic period around 7000 BC (Malinowski and Garfinkel 1991). The *Ten Books on Architecture* by Vitruvius written in the latter part of the first century BC describe how calcined limestone hardens when mixed with water and also describes the hydraulic setting properties of natural pozzolans. However, the prototype of the modern cement used today was invented in Britain in 1824 by J. Aspdin.

Since then, the consumption of cement concrete has grown enormously, and has been particularly remarkable in the last half century. Figure 1.1 shows the volume of cement and steel production, together with the changes in population and per capita GDP since 1950 by setting the 1960 levels at a value of 100. While steel production has approximately tripled in the last fifty years, cement production has increased approximately 9.5 times. At present, although no definitive data is available, total concrete production volume is thought to be around 20 billion tonnes. It is interesting to note that the growth in per capita GDP is twice the growth in cement production volume. This demonstrates how great the role of infrastructural investment is in creating wealth.

The reasons for the great rise in the use of concrete over the last half century are simple. The first is that approximately 70% of the material for concrete consists of the rocks most commonly found on Earth. The second is that the limestone and clay that are the raw materials of cement production are also present in great abundance. Water of course needs no explanation. To build a more convenient and affluent society, economic development is essential. This inevitably leads to the development of socioeconomic infrastructure, the basic material for which is concrete.

With the collapse of the controlled economies of the socialist countries, the quest for more affluent lifestyles has accelerated development of the free market economy. The majority of the world's population is found in the BRICS and other emerging and developing countries. As these countries aspire to a standard of living equal to those of developed countries, they are driving consumer activity. To create an affluent society, construction investment is vital.

Figure 1.2 shows the relationship of per capita GDP to construction investment in major nations. It is clear that the growth of construction investment raises GDP. Of the developed countries, Germany and Japan appear to have

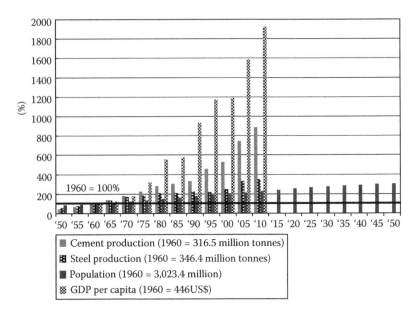

FIGURE 1.1
Cement and steel production after 1950.

followed the same trajectory, and the sharp increase in German construction investment appears to be due to the reunification of East and West Germany. Construction investment and GDP in the United States, France, and Britain show very clear proportional relations. On the other hand, while China is undertaking rapid construction investment, it only amounts to around a third of the investments of developed countries and its GDP is only one seventh. Meanwhile, in India and Vietnam that appear to be in remarkable growth phases, the levels of construction investment and GDP are conspicuously low. Going forward, it is unclear what development path China and India, by far the most populous countries in the world, will follow, and it is possible that they will adopt different paths from those of developed countries. However, given the current cement and steel production levels as shown in Figure 1.3 and Figure 1.4, it is predicted that the volume of cement and steel production in China and India will experience major growth and that their share in the world total will become very significant.

There are two problems concerning concrete and steel that are necessary for the development of socioeconomic infrastructures. One is that despite abundant availability of their material resources, the volume of consumption is enormous. Another is that the production of cement and steel generates a large amount of carbon dioxide, known to be a "greenhouse" gas. However, there are no alternatives to concrete and steel as basic materials for the development of socioeconomic infrastructure.

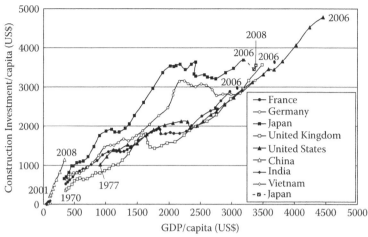

Japanese figures for 2007 and 2008 calculated from construction investment figures of the Research Institute of Construction and Economy.
China/India/Vietnam: IMF (2001–2008),
United States: OECD (1977–2006), other: OECD (1970–2006).

FIGURE 1.2
GDP and construction investment by major nations.

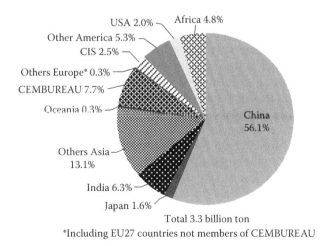

Total 3.3 billion ton
*Including EU27 countries not members of CEMBUREAU

FIGURE 1.3
World cement and steel production. (*Source:* CEMBUREAU (European Cement Association). 2010. *Activity Report*. Brussels.)

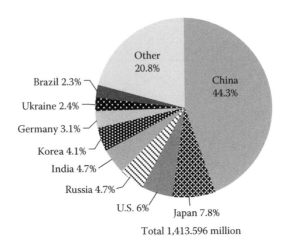

FIGURE 1.4
World cement and steel production. (Source: World Steel Association. 2010. Brussels. http://
www.worldsteel.org/)

The scope for the use of timber is very limited. This is clear when we consider that building timber structures for the world's 7.0 billion people to live
and work in would require huge increases in land use and supplying the
timber would cause serious environmental problems. Mankind thus has no
choice but to continue utilizing concrete even though significant increases
in material consumption and the resulting carbon dioxide emissions due to
growing population and socioeconomic expansion will most likely become
strong factors in hampering sustainable global development. Moreover, these
trends will be accompanied by the emergence of socioeconomic problems,
the resolving of which is important in order for the concrete technology system currently in use to shift to a system that allows for the sustainable use of
concrete. For this purpose, current technology needs to be assessed from the
perspective of sustainability.

This book begins by clarifying the historic background and meaning
of sustainability, after which it outlines areas that need to be considered
in connection with sustainability in the concrete and construction field.
An outline is also given of the currently available methods of assessing
sustainability. There follows an explanation of sustainable technologies
related to concrete and concrete structures, while many examples are presented to promote understanding of the current state of such technologies. Finally, the authors look ahead to the future we should aim for in the
twenty-first century.

References

CEMBUREAU (European Cement Association). 2010. *Activity Report*. Brussels.

Davidovits, J. 1987. Ancient and modern concretes: what is the real difference? *ACI Concrete International*, 9, pp. 23–35.

Malinowski, R. 1979. Concrete and mortars in ancient aqueducts. *ACI Concrete International*, 1, pp. 66–76.

Malinowski, R. and Garfinkel, Y. 1991. Prehistory of concrete. *ACI Concrete International*, 13, pp. 62–68.

World Steel Association. 2010. http://www.worldsteel.org/

2

Sustainability

2.1 Natural History

It is assumed that the Earth is 4.6 billion years old and that life evolved about 4 billion years ago. Since then, the Earth repeatedly experienced climate warming and cooling. This is known to have arisen through a complex process involving multiple factors including changes in gases in the atmosphere caused by major volcanic activity and photosynthesis, asteroid collisions, changes in the inclination of the Earth's axis of rotation, the eccentricity of its orbit, and its precession. On a geological time scale, the global environment has experienced carbon cycling, in the process of which carbon fixation has occurred.

Varied levels of the greenhouse effect according to the concentrations of carbon dioxide in the atmosphere have exerted extensive impacts on global climate. The Earth has reached its current state through such immensely complex processes. Although Venus is thought to have a similar formation and composition, its surface temperature exceeds 450°C because of its proximity to the Sun and its atmosphere that consists mostly of carbon dioxide and in fact produces the ultimate greenhouse effect. This extreme greenhouse effect is evident when we consider that the average surface temperature of Mercury, which is the closest planet to the Sun but has little atmosphere, is only 179°C. On the other hand, although Mars is also enveloped by an atmosphere containing carbon dioxide, its average surface temperature is −43°C, as the concentration of carbon dioxide is very low (a smaller greenhouse effect). Situated between Venus and Mars, the Earth has a surface consisting mostly of nitrogen and oxygen, while the carbon dioxide level is about 0.038%. The global warming we are currently facing is due to the vast discharges of carbon dioxide within a short period of time, following the consumption of oil, coal, and limestone to which carbon has fixed over a long period. In other words, humankind is partly responsible for accelerating global carbon cycling. It is evident that precisely predicting the level of the global greenhouse effect is next to impossible. However, considering the foregoing facts holistically, it is not difficult to assume that to continually release fixed carbon in the form of carbon dioxide into the atmosphere represents a gradual return to the early stage of this planet when its atmosphere consisted mostly of carbon dioxide.

2.2 Human Beings and Concrete

On the Earth that is 4.6 billion years old, early humans and *Homo sapiens* emerged 5 million and 3 million years ago, respectively. Modern humans originated in Africa approximately 100,000 years ago, and their descendants migrated throughout the world. Agricultural society was formed only 10,000 years ago, but the invention of agriculture significantly increased the global population that had formerly been no more than 50 million. During the Roman period lasting roughly from 500 BC to 500 AD, the world population increased to 200 million. After the next 1,000 years (the Middle Ages), the Industrial Revolution occurred in England in the middle of the eighteenth century, when the population is said to have reached 800 million.

Now, 250 years later, the global population is 7.0 billion. In other words, it is 35 times the population during the Roman period, and more than 8 times that at the time of the Industrial Revolution. The global population is predicted to increase further in the future to 9 to 10 billion and most of the increases will be in developing nations. This suggests that the consumption of resources, energy, and food will drastically increase, and it is highly possible that consumption will exceed the environmental capacity of a celestial sphere with a radius of a little more than 6,300 km.

Although the origin of cement extends back to the Roman period, fewer than 200 years have passed since modern cement was invented, and it was only half a century ago that concrete began to play a central role as a construction material. Since then, concrete technology has shown dramatic development, and combined with a rising population and economic growth, its consumption has augmented rapidly and is now the material most used for construction (second only to water) worldwide. The projected increase in concrete consumption will have a great impact on the global environment. Another predicted factor, in addition to the consumption of resources, is that the emissions of CO_2 arising during cement and steel production will also account for a considerable percentage of its total emissions.

2.3 Genealogy of Sustainability

More than 100 years ago, in 1895, three British philanthropists founded the National Trust with the aim of protecting nature and buildings of historic value against unrestrained development. The Britain of those days had been the scene of continuous rapid economic development since the Industrial Revolution of the mid-eighteenth century. It became known as the "workshop of the world" and indiscriminate development and industrialization caused severe problems. The organization these three men founded continues to this day with its activity of acquiring and preserving precious stretches of coastline, areas of countryside, and buildings. The organization's activity deserves recognition as the first time humans acknowledged the value of the environment and took systematic and concrete action to ensure its continuance.

In 1972, at a time when the world was on the brink of a period of rapid economic development, a report titled *The Limits to Growth* (Meadows et al., 1972) was presented to the Club of Rome. The report concluded that if population growth and environmental pollution continued, growth on the planet would reach its limit within 100 years. This conclusion was shocking, but because of the timespan involved it was not received with great seriousness.

The first conference at which "the environment" was discussed at the international level was the United Nations Conference on the Human Environment held in 1972 in Stockholm. The Declaration of the United Nations Conference on the Human Environment was adopted, and the following principle regarding human rights and duties relating to the environment was proclaimed:

> Man has the fundamental right to freedom, equality and adequate conditions of life, in an environment of a quality that permits a life of dignity and well-being, and he bears a solemn responsibility to protect and improve the environment for present and future generations.

The United Nations Environment Programme (UNEP) was established as the organ to implement this environmental declaration. In 1987, the UN World Commission on Environment and Development published the so-called Brundtland Report which for the first time gave an outline definition of the well-known concept of sustainable development:

> Sustainable development is development that meets the needs of the present without compromising the ability of future generations to meet their own needs.

The report also states clearly that the goals of economic and social development should be determined from the perspective of sustainability, but a clear definition of sustainability is not provided. However, the report gives sustainability the status of an economic and social system and states that economy and ecology must be completely integrated. In other words, sustainability can be seen as resting on the three pillars of economy, society, and the environment, which is the general interpretation made nowadays. A further noteworthy point of the report is the following prescient statement on global warming:

> The burning of fossil fuels puts into the atmosphere carbon dioxide, which is causing gradual global warming. This 'greenhouse effect' may by early next century have increased average global temperatures enough to shift agricultural production areas, raise sea levels to flood coastal cities, and disrupt national economies.

The report can thus be seen as a very advanced document that states with great clarity the existence of the problems we face today and predicts their future trends. In 1992, twenty years after the United Nations Conference on the Human Environment, the UN Conference on Environment and Development (the Earth Summit) was held in Rio de Janeiro and adopted the Rio Declaration. The fourth principle of the declaration makes the following statement:

> In order to achieve sustainable development, environmental protection shall constitute an integral part of the development process and cannot be considered in isolation from it.

This step again emphasized the importance of sustainable development to combine environmental conservation with economic development.

The Earth Summit adopted the United Nations Framework Convention on Climate Change that came into force in 1994. Based on the principle of the "common but differentiated responsibilities" of the signatory nations, the convention placed on developed signatory nations a duty to implement measures for greenhouse gas reduction. Every year since 1995, a meeting of the signatory nations known as the Conference of the Parties (COP) has taken place. In 1997, COP3, convened in Kyoto, adopted the Kyoto Protocol that went into effect in 2005 and set greenhouse gas reduction targets.

As well as setting a greenhouse gas reduction rate for each country in the form of a target based on 1990 levels to be achieved by 2008 through 2012, the Kyoto Protocol attempted to further facilitate greenhouse gas reduction by introducing the so-called Kyoto mechanisms (clean development, emissions trading, and joint implementation). At COP15 (2009) and COP16 (2010), discussions were held on a post-Kyoto Protocol to consist of greenhouse gas reduction targets for the period after 2013, but a sharp division among the developed countries claimed that there was no point in setting targets that did not include the developing countries. The developing nations claimed that the developed countries caused the current problems; the impasse has continued. The 35th G8 Summit (2009) reconfirmed the target of a reduction of at least 50% in total worldwide greenhouse gas emissions by 2050, as part of which backing was given to a target reduction of at least 80% by 2050 for the developed countries as a whole.

In 1988, UNEP and the World Meteorological Association set up the Intergovernmental Panel on Climate Change (IPCC) that has to date published four periodic assessment reports. The fourth report, published in 2007, stated explicitly that global warming is an established fact and is caused by human activity. Table 2.1 shows the required emission levels for different groups of stabilization concentrations and the resulting equilibrium global warming and long-term sea level rise due to thermal expansion only.

Current atmospheric CO_2 concentration exceeds 380 ppm, but it is fair to say that common agreement has been reached on the Category I target of limiting temperature rise to around 2°C.

The IPCC's fourth assessment report presents examples of global warming mitigation technology in major sectors of industry. Examples given for the construction sector include a modal shift from road transport to rail and public transport systems, integrated building design using technology that allows feedback and control, improved energy efficiency in the cement and steel industries, introduction of carbon capture and storage (CCS) technology, and recycling and waste reduction. Currently, work on the fifth assessment report has begun and publication is due in 2015.

2.4 Pillars of Sustainability

2.4.1 General

Humankind has created living and production environments for itself using the limited resources available on planet Earth. The consumption of resources has increased in line with the growth in the global population and a range of issues has arisen in connection with the process of consumption. These issues may now have reached a stage where they have passed a critical point in the Earth's environmental carrying capacity. If the Earth's environment is understood as an ecosystem, it can be defined as the interactions of living organisms, dead organic matter, and abiotic environments (soils, water, and atmosphere). It is vital that the balance of this system should not be disturbed to the extent where it passes the point of recovery.

Mankind has over a long period developed a system of division of labor and established a wide range of industries. All industries basically make use of global resources and therefore can be said to exist based upon the imposition of some kind of burden on the environment. The concrete and construction sector is no exception. The paradox is thus that the modification of the environment to create human amenity actually depends on environmental degradation. However, humankind has no other option but to minimize this environmental degradation and thereby work to maximize human happiness. What we therefore need to do is to identify the issues as clearly as

TABLE 2.1

CO$_2$ Concentrations and Effects Cited in IPCC's Fourth Report

Category	CO$_2$ Concentration at Stabilization (2005 = 379 ppm) (ppm)	CO$_2$ Equivalent Concentration at Stabilization Including GHGs and Aerosols (2005 = 375 ppm) (ppm)	Peaking Year for CO$_2$ Emissions (years)	Change in Global CO$_2$ Emissions in 2050 (percent of 2000 Emissions) (percent)	Global Average Temperature Increase above Preindustrial at Equilibrium, Using "Best Estimate" Climate Sensitivity (degrees C)	Global Average Sea Level Rise above Preindustrial at Equilibrium from Thermal Expansion only (meters)	Number of Assessed Scenarios
I	350 to 400	445 to 490	2000 to 2015	−85 to −50	2.0 to 2.4	0.4 to 1.4	6
II	400 to 440	490 to 535	2000 to 2020	−60 to −30	2.4 to 2.8	0.5 to 1.7	18
III	440 to 485	535 to 590	2010 to 2030	−30 to +5	2.8 to 3.2	0.6 to 1.9	21
IV	485 to 570	590 to 710	2020 to 2060	+10 to +60	3.2 to 4.0	0.6 to 2.4	118
V	570 to 660	710 to 855	2050 to 2080	+25 to +85	4.0 to 4.9	0.8 to 2.9	9
VI	660 to 790	855 to 1130	2060 to 2090	+90 to +140	4.9 to 6.1	1.0 to 3.7	5

possible and take action to resolve them. This section gives an outline of the problems that exist with reference to the three pillars of sustainability and examines their relationship to the concrete and construction industry.

2.4.2 Environmental Issues

Environmental issues can be conveniently identified with reference to the spatial scale indicated below and shown schematically in Figure 2.1:

- Global environment
- Regional environment
- Local environment
- Built environment

Issues concerning the global environment involve biodiversity, global warming, ozone destruction, and resource depletion. Biodiversity and resource depletion are phenomena that appear locally and regionally, but biodiversity is directly connected to the global system, and as resources are nowadays transported and used on a global scale, it is appropriate to deal with these as global-level environmental issues.

All these issues are relevant to the concrete and construction sector. The extraction of construction materials and construction activities lead directly to the destruction of nature and have big impacts on natural habitats. Cement and steel manufacture causes the emission of large volumes of CO_2 and construction activity uses large amounts of fossil fuel. These activities contribute to global warming. The Earth's atmospheric ozone layer absorbs the Sun's harmful ultraviolet rays, and is thought to thus play the role of protecting the Earth's ecosystem. However, it has been discovered that chlorofluorocarbons and other chlorine-based chemical substances destroy ozone.

In the construction sector, chlorofluorocarbons have been used in urethane foam insulating materials and in air-conditioning and heating systems.

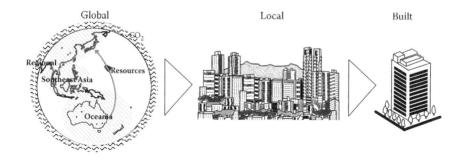

FIGURE 2.1
Environmental issue classifications

Since the adoption of the Montreal Protocol, regulation has become stricter, and chlorofluorocarbon substitutes are now used. However, the buildings that existed then contain now restricted materials, and thus appropriate measures are required when they are demolished. Rapid increases in the consumption of resources in developing countries results in their massive exploitation. This not only leads to a global-scale problem of resource depletion, but also to a destruction of the natural environment, and ultimately impacts the global system.

The regional environment is defined as a country or a wider environmental segment. At this environmental level, issues such as acid rain, air pollution, and water pollution occur. In the manufacture of a range of products, raw materials and fuel or energy are used. Many of these contain naturally occurring harmful substances; alternatively, harmful substances may be produced in the manufacturing process. It is important to recover these substances appropriately, but the relevant technologies and costs are obstacles, and emissions may be released without always undergoing appropriate treatment. This is a problem particularly likely to arise in developing countries.

The resulting acid rain, air pollution, and water pollution can affect a considerable area, so that even if a country does not generate such problems, it may be impacted negatively by production activity in other countries or regions. Relevant issues for the concrete and construction sector are the types of pollution associated with steel and cement manufacture and with the thermal power generation that provides its energy source.

The local environment is defined as a city or other specific area of the environment. At this environmental level, issues such as soil contamination, waste generation, land use, the heat island effect, noise, vibration, and dust occur. These issues can be observed and experienced within the scope of our daily lives. In the concrete and construction sector, soil contamination and waste generation associated with the use of concrete result. Land use is also essential to construction. Because it alters the natural environment, land use impacts the natural habitats of living creatures. Cities built with concrete are well known to be subjected to the heat island effect. Moreover, a construction site generates noise, vibration, and dust over a relatively extensive area, thus impacting neighboring localities. When a structure is demolished, a great amount of waste is generated, and disposal and recycling treatment also impose environmental burdens at the local and global environment levels.

The built environment is defined as the living and working environment and its surroundings. Buildings forming the living and working environment are subject to indoor pollution caused by volatile organic compounds (VOCs), fungi, and radiation. The construction of buildings destroys the surrounding natural environment, although revegetation can effect a partial recovery. High-rise buildings cause the wind tunnel effect. They also impede wind circulation by blocking cool winds from the mountains and the sea in summer and therefore contribute to temperature rises in cities. The construction of buildings may also impair harmonization with the surrounding scenery. On the other hand, the concrete used in buildings may have a positive impact through the thermal mass effect that reduces the energy required to operate a building. Additionally, high-rise buildings can reduce land use.

2.4.3 Economic Issues

Environmental issues arise from industrial production and the provision of various services that are designed to add to the convenience and richness of human life. In other words, economic activity that uses large amounts of resources and energy is the root cause. Contemporary society engages in production activity based on the assumption of economic growth, which in turn assumes constantly rising consumption of resources and energy; any suspension of this growth is described as economic recession. The developed countries already have a rich supply of basic goods, the renewal of which supports the economy. The developing countries, however, have insufficient goods, which has led to the expansion of production activity and sustained rapid economic growth outside those countries. Developing countries are important markets for the developed world as they expand exports and production. The world economy can be said to operate based on such complex economic mechanisms.

What makes this economic activity possible is the provision of what is called infrastructure. Providing infrastructure is the job of the construction industry. In specific terms, this means the construction of buildings, roads, railroads, seaports and airports, bridges, tunnels, dams, and the other structures that drive socioeconomic activity, as shown in Figure 2.2. Buildings are required for the purpose of supervising and otherwise managing the production of goods and economic and social activity. In modern society, roads are the bases for the most important means of goods transport and human movement. Seaports are essential for the mass transport of resources and products and are used as bases for passenger ship services.

Airports are of growing importance as facilities for take-off and landing of aircraft carrying passengers and freight. Today, the number of airports around the world is more than 8,700. Approximately 15,000 aircraft are in service worldwide, a figure thought likely to double. Dams, which are sometimes spoken of as symbols of environmental destruction, have a range of functions covering water supply, flood control, and electricity generation. Electricity generation by dams produces almost no CO_2. Hydroelectric generation accounts for 100% of electricity production in Norway and 70% in Brazil. Japan has more than 2,800 dams, but the proportion of electricity generated by them is no more than about 7%.

From the above, it can be concluded that the role of the construction industry in human economic activity is very great, but as massive amounts of

Buildings in Tokyo (Oedo 21)   Hoover Dam and Bridge (Obayshi Corp.)   Tokyo International Airport (Taisei Corp.)

Dubai Monorail (Obayashi Corp.)   Millau Viaduct (CEVM/Daniel Jamme)

FIGURE 2.2
Infrastructures that drive socioeconomic activity.

resources and energy are used in infrastructural provision itself, the costs are also enormous. In order to minimize the investment required for construction, it is therefore important when infrastructure is created to consider the economic benefit also after ensuring the necessary function and performance. Caution is nevertheless needed, as infrastructure in general has a high degree of public benefit and excessive emphasis on economic benefit will increase risks to safety.

Infrastructure also needs to be considered in terms of the economic benefit over its life cycle. As major repair or strengthening of concrete structures during their useful life is undesirable, increasing the initial cost in order to reduce the life cycle cost may be one option. Construction investment has a wide impact and a large multiplier effect in the macroeconomy. Infrastructure brings huge benefits through economic activity.

There is a very delicate relationship between the economy and the environment. Until now, the development of the economy has meant increased consumption of resources and energy, resulting in environmental degradation. How to overcome this dilemma through the concept of sustainability is the biggest task facing mankind today. One idea from the perspective of resource consumption is "producing more with less." Another key is repeated use of the resources consumed to minimize further extraction. Ultimately, it may be necessary to limit the global population to an appropriate level.

There are several approaches through which the concrete and construction sector can contribute to resolving these problems. First, it can use technological innovation to drastically reduce the environmental burden imposed by the materials used in construction; the authors will return to this question later. Second, it can improve the quality of the infrastructure in order to create a new social system, in other words, realize infrastructure development to facilitate a social system that minimizes the environmental burden and the cost of socioeconomic activity. This does not necessarily mean a departure from a contemporary motorized society. What is important is to achieve a major reduction in the environmental burden and cost of human movement and transportation of goods. Facilitating full recycling of the concrete which is a major resource of the construction sector is also very important in efforts to reduce the consumption of virgin resources. In this connection, another construction material, steel, is already almost 100% recycled. This leads to reductions of cost and of CO_2 emissions.

2.4.4 Social Issues

All human activity involves a social aspect of some kind. Environmental and economic issues involve a large number of social factors. The classic example is the economic gap between developed and developing countries that gives rise to political conflict and tense relations. Poor social systems cause environmental issues that become social problems. Developing countries, for whom economic growth is an absolute imperative, need to work proactively on infrastructural development. Infrastructural development creates the basis for economic development, but at the same time of course it greatly increases the environmental burden caused by economic development. Our ultimate goal should be a peaceful world with an equal distribution of wealth, enjoying a healthy global environment. Achieving these conditions is not easy, but we need to consider holistically the three elements for achieving sustainability.

The social value of infrastructural development is varied in character, but broadly speaking it would include the following aspects, and the degree to which these are achieved is a social issue:

1. Quality (as a place to live and work in a building)
2. Safety and security
3. Serviceability
4. Aesthetics
5. Efficient land use
6. Protection of cultural heritage
7. Protection of nature from dangerous substances
8. Job opportunities

The varied characteristics of concrete make it possible to create a variety of concrete structures corresponding to differing social needs. If the use of a structure is limited due to reduced durability of the concrete, this has a large social impact. Structures with low seismic resistance increase the damage caused by earthquakes, with loss of human life as the greatest negative social impact.

The aesthetic value of concrete structures has a large social impact in terms of its relationship with the surrounding scenery. The building of concrete structures destroys ecosystems through land use, but can on the other hand minimize destruction. Additionally, quality concrete and concrete structures facilitate economic activity, protect cultural assets, and also have the function of protecting nature from the spread of hazardous substances. Concrete structures also provide spaces for cultural activity. The construction industry, which uses concrete as one of its principal construction materials, is a very large industry, and supports a large number of jobs. Infrastructural development has long been seen as having the twin benefits of creating the basis for socioeconomic activity and creating employment.

As we have seen, concrete is an important material in building an environment that can support stable socioeconomic activity and its social role is therefore very great. On the other hand, as it necessitates the use of conspicuously large amounts of resources and energy, it also involves a great environmental burden. We need to engage seriously with the question of how to achieve sustainability in construction.

References

Intergovernmental Panel on Climate Change. 2007. Climate Change. Synthesis Report Summary for Policymakers.

Meadows, D.H., Meadows, D.L., Randers, J. et al. 1972. *The Limits to Growth*. Universe Books, Singapore.

World Commission on Environment and Development. 1987. *Our Common Future*. Oxford University Press, Oxford.

3

Sustainability in Concrete and Construction

3.1 Environmental Aspects

3.1.1 Materials Flow, Resource Consumption, and Waste Generation

3.1.1.1 Materials Flow

Advanced industrial nations enjoy a convenient and affluent lifestyle through the production and consumption of various products and services, using a vast amount of resources extracted from the natural environment. However, incidental to such production and consumption, large amounts of waste gas, fluid, and other products are generated, which result in waste solids and eventually return to the natural environment such as the atmosphere, hydrosphere, and geosphere.

Various environmental issues are basically attributable to the fact that a material flow consisting of such vast extraction of resources, production and consumption of products and services, and generation of wastes has largely exceeded the capacity of the natural environment in terms of resource regeneration and waste purification. In order to achieve sustainable development, it is essential for us to acknowledge that the availability of resources and purification capacity of the natural environment are limited.

Chapter 4: Changing Consumption Patterns of Agenda 21 adopted at the Earth Summit (UNCED) in 1992 notes that the production and consumption patterns of developed industrial nations are no longer sustainable. Although resource input generally increases with a growing economy, it is an important factor for achieving sustainable development to keep the increasing rate of resource input lower than that of economic growth, in other words, to optimize resource productivity (GDP/resource input). It is therefore important to conduct a quantitative analysis of the relation between the natural environment and human society, while understanding and at the same time predicting current and future material flows respectively, in order to create and maintain a society based on sustainable development in the future by solving the environmental issues of today's mass-production and mass-consumption society.

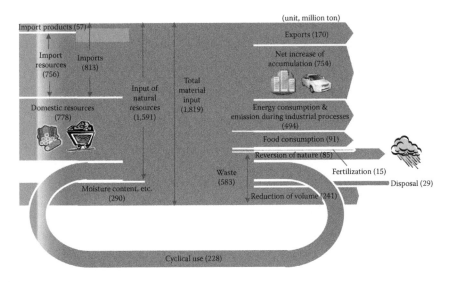

FIGURE 3.1
Overview of Japan 2006 material flows. (*Source:* Ministry of the Environment, Government of Japan. 2009. Establishing a sound material-cycle society, creating economic development through the establishment of a sound material-cycle society. With permission.)

Figure 3.1 shows Japan's 2006 material flow, providing an overview indicating that total material input was 1.82 billion tonnes, of which 40%, i.e., 750 million tonnes were accumulated in the form of buildings and civil engineering structures. One hundred and seventy million tonnes were exported in the form of products, and 490 million tonnes were consumed as energy or emitted into the natural environment during industrial processes, while 580 million tonnes were generated as waste. It is revealed that only 230 million tonnes, one-eighth of the total input of materials were for cyclical use. This also shows that, as the amount of ready-mixed concrete produced in 2005 was 118,981,736m³ (approx. 273 million tonnes), 15% of the material input was for the production of concrete. Considering that the demands for other major building materials such as steel (crude steel output) and timber (including those for plywood and woodchips) in the same period in Japan were approximately 71 million and 15 million tonnes respectively, the concrete production volume was overwhelmingly high.

Around the year 2000 when the construction sector was thriving in Japan, as much as approximately 25% of the materials were used for the production of concrete. There is no doubt that all nations, not only western countries or Japan, without exception, experience the same situation regarding material input for concrete production during the course of their advancement through the development of their social infrastructure. It is known that concrete is the second most consumed substance on Earth after water (Cohen et al., 2006). This demonstrates that concrete has an enormous impact on the creation of a society with sustainable development and a recycling-oriented

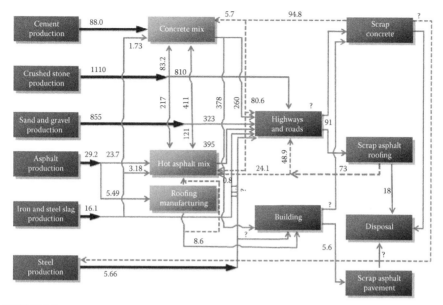

FIGURE 3.2
United States 1996 material flows. (*Source:* Kelly, T. 1998. Crushed Cement Concrete Substitution for Construction Aggregates—A Materials Flow Analysis, U.S. Geological Survey Circular 117.7. With permission.)

society, not only from the aspect of the consumption of resources, but also production and future waste generation.

Next, we focus on the material flow related only to concrete. Figure 3.2 and Figure 3.3 show the material flows of concrete and asphalt concrete in the United States (1996) and Japan (2005), respectively. To produce 1 m³ of concrete, approximately 250 to 350 kg of cement, 150 to 200 kg of water, and 1,800 to 2,000 kg of aggregate (60 to 75% of concrete volume) are required. As limestone and clay are the main raw materials for cement, concrete production involves the extraction of a vast amount of rock minerals from the natural environment and their mechanical and thermal processing.

In addition, as concrete can be produced in an enormous volume and be mixed with inorganic minerals without degrading its performance, various inorganic wastes from other industries including blast furnace slag and fly ash are largely used as raw materials for the production of cement and concrete. Such enormous production volume means that the energy required for concrete production and transportation is also enormous. Thus it is essential to understand the material flow of concrete, not merely from the standpoint of resource saving, but also with a view toward energy saving.

3.1.1.2 *Resource Consumption*

While it is evident that the Earth's resources are not infinite, they nevertheless represent a huge amount when compared with the volume that humankind

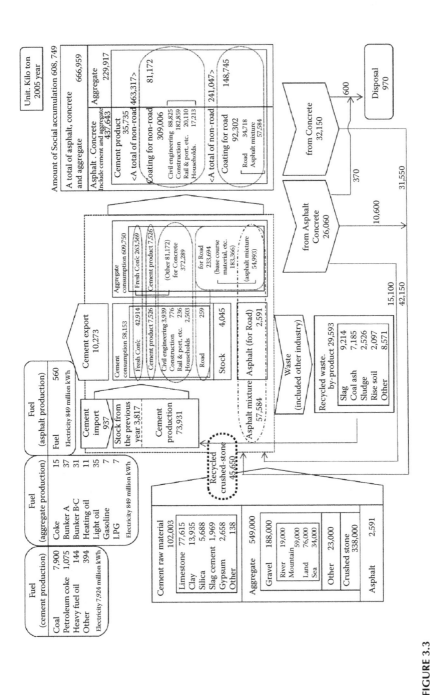

FIGURE 3.3

Japan 2005 material flows. (*Source:* Committee on Resource Utilization Strategy, Architectural Institute of Japan. 2008. Challenge of Material Flow and Resource Utilization in Building Materials. With permission.)

uses today. However, the resources we can actually put to use are limited, and we should acknowledge that they will eventually be depleted. In other words, depletion of resources does not mean that they will totally disappear, but fundamentally means:

- Significant reduction in availability to the next generation
- Difficulty in acquiring resources as a vast amount of energy is needed to do so
- Difficulty in carrying out life activities that depend on resources

In real terms, depletion of resources coincides with the following phenomena in the acquisition of a given resource:

- Soaring resource price
- Increased environmental destruction through acquisition of resource
- Loss of economic advantage regarding the use of resource

These exhaustible resources include fossil fuels (see Section 3.1.1.3), minerals, soil, and stones. Although the reserve of soil and stones existing in the Earth's crust is bountiful compared with fossil fuels and minerals, one cannot say it is inexhaustible when we consider that the extractable amount is controlled by laws related to environmental protection. Further, the specificity of natural resources, i.e., finiteness and maldistribution, destabilize resource supply, causing incidental environmental and economic problems to future generations.

Taking mineral resources as an example, in principle it is technically feasible to continue mining them by accepting lower grades, however, in such a case, the amount of energy required for ore dressing and refining and the wastes generated will be enormous. Therefore lower grades cannot serve as economical resources.

Depletion can become a serious issue when the economically mineable reserves of resources are scarce and at the same time their consumption is increasing. The mineable volume of such resources could increase with the improvement of mining technology in the future, but if not, they will become unobtainable due to economic constraints. However, if a high grade recycling scheme, namely closed loop recycling is thoroughly implemented, the depletion of natural resources will not present such a big problem.

Figure 3.4 shows changes in resource input in the United States. Despite some impacts from oil shocks and economic depressions after World War II, the input continued to rise sharply. It was 144 million tonnes in 1900 and 3.78 billion tonnes in 2006, showing an increase of 26 times over a period of more than 100 years; 77% or 2.91 billion tonnes were used for construction materials. In Europe, as shown in Figure 3.5, the resource input was between 7 billion and 8 billion tonnes after the year 2000, and as Figure 3.6 suggests, more than half the resources were likely to have been used for construction materials.

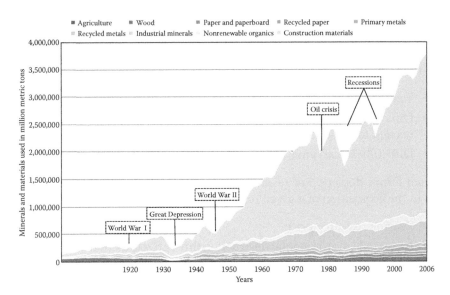

FIGURE 3.4
United States resource input trends. (*Source:* Matos, G. R. 2009. Use of Minerals and Materials in the United States from 1900 through 2006. U.S. Geological Survey Fact Sheet 2009-3008. With permission.)

However, in Europe, the production and consumption of resources differs greatly depending on the country, as shown in Figure 3.7. There are countries with high import and export ratios such as Belgium and the Netherlands, those that have established local production for local consumption such as

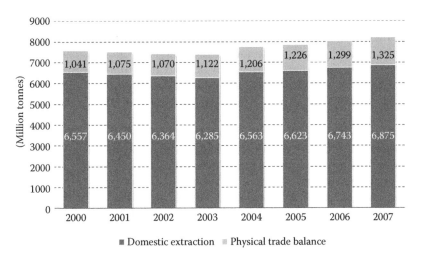

FIGURE 3.5
Europe resource input trends. (*Source:* Eurostat, European Union. 2010. Environmental statistics and accounts in Europe. With permission.)

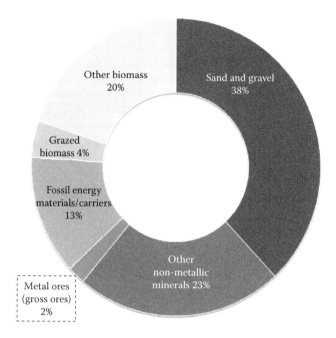

FIGURE 3.6
Materials extracted and used In EU 27 countries, 2007. (*Source:* Eurostat, European Union. 2010. Environmental statistics and accounts in Europe. With permission.)

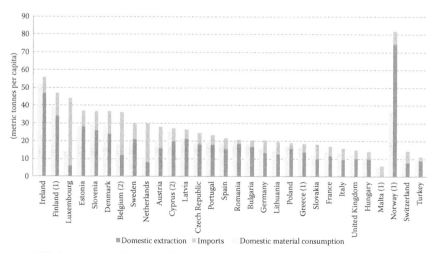

(1) Trade data are estimated using external trade statistics.
(2) Estimates.

FIGURE 3.7
Production and consumption of resources by country. (*Source:* Eurostat, European Union. 2010. Environmental statistics and accounts in Europe. With permission.)

TABLE 3.1

Cement and Aggregate Consumption and Concrete Production in the United States, Europe, and Japan (million tonnes)

Country and Year	Cement Consumption (A)	Aggregate Production (B)	Ready-Mixed Concrete Production (C)	Aggregate/ Cement Ratio (B/A)	Ready-Mixed Concrete/ Cement Ratio (C/A)
U.S. 2008	127.3[a]	Crushed stone : 1,440[b] Gravel : 1,040[c]	270[a]	19.5	2.12
U.K. 2006	12.1[d]	274[e] Crushed stone : 123	25[a]	22.6	2.07
France 2006	22.3[d]	430[e] Crushed stone : 233	43[a]	19.3	1.93
Germany 2006	33.5[d]	625[e] Crushed stone : 270	43[a]	18.7	1.28
Italy 2006	47.8[d]	354[e] Crushed stone : 135	78[a]	7.4	1.63
Spain 2006	54.0[d]	486[e] Crushed stone : 314	98[*1]	9.0	1.81
Japan 2009	49.2[f]	456[g] Crushed stone : 299	86	9.3	1.75
World 2009	3,000	45,300	5,400	Average 15.1	Average 1.80

[a] http://www.ermco.eu/documents/ermco-documents/statistics.xml?lang=en
[b] http://minerals.usgs.gov/minerals/pubs/commodity/stone_crushed/mcs-2010-stonc.pdf
[c] http://minerals.usgs.gov/minerals/pubs/commodity/sand_&_gravel_construction/mybl-2008-sandc.pdf
[d] http://www.indexmundi.com/minerals/
[e] Production of aggregates in European Union.
[f] http://www.jcassoc.or.jp/cement/3pdf/jh3_0500_d.pdf
[g] http://www.saiseki.or.jp/pdf/20shinzai.pdf

Bulgaria, Cyprus, Finland, Ireland, Hungary, Poland, and Rumania, and those that are natural resource exporting countries such as Norway.

Although there is no statistical data available regarding the worldwide production volume of concrete, its current and future levels can be estimated based on the current and future estimated usage of its constituent materials and the ratios of concrete production volume to its constituent materials used in a specific region. Table 3.1 shows the volumes of cement and aggregate consumed and ready-mixed concrete produced in the United States, European countries, and Japan, respectively. The ratios of cement

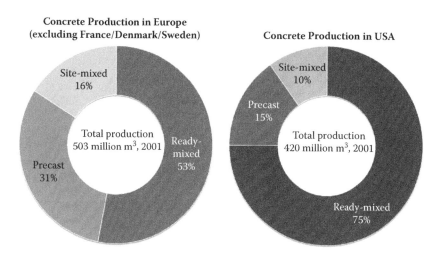

FIGURE 3.8
Comparison of types of concrete produced in Europe and United States.

to aggregate used and to ready-mixed concrete produced are 16.1 and 1.76 times, respectively. However it should be noted that concrete is produced as ready-mixed concrete, site-mixed concrete, and precast concrete products. Figure 3.8 shows the production types of concrete in Europe (ERMCO, 2002) and the United States (U.S. Geological Survey, 2002). It reveals that Europe has a lower production percentage in ready-mixed concrete, and higher on site-mixed and precast concrete compared with the United States. Assuming that ready-mixed concrete accounts for two thirds of total concrete production, worldwide volume is estimated to be 8 billion square meters.

Global cement production in 2009 was approximately 3 billion tonnes, of which more than two thirds was produced in Asia as shown in Figure 3.9. In China particularly, cement accounted for more than 50% of the entire world production, and in India also it is expected to grow dramatically in line with the country's soaring economic growth. With respect to the future trend of global cement production, prediction was made in 2002 concerning its demand based on the four scenario "story lines" regarding future global socio-economic growth and increases in the world's population. The story lines were developed as part of IPCC activities (IPCC, 2000) (Table 3.2). As shown in Figure 3.10, in the case of storyline A1, global demand for cement in 2020 is predicted to be approximately 3.7 billion tonnes, which is 180% higher than demand in 1990 (approximately 1.3 billion tonnes) and even higher (5.4 billion tones) in 2050.

The key factor for this is without doubt attributable to projected increases in developing nations. Figure 3.11 shows the predicted amounts of mineable reserves of limestone—the key raw material of cement—in Japan (confirmed, confirmed + estimated, confirmed + estimated + expected) and cumulative

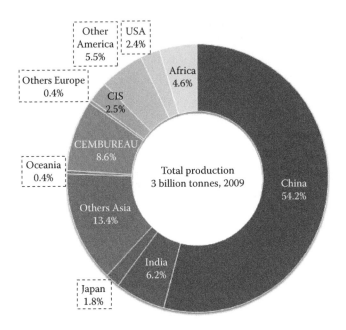

FIGURE 3.9
Global cement production. (*Source:* CEMBUREAU. 2009. Activity Report 2009. With permission.)

consumption (including consumption as raw material for cement and aggregates and also for the refining of iron and steel). This demonstrates that mineable limestone will be depleted by about 2050 in Japan, which is not hard to anticipate, considering the increasing global cement production.

Figure 3.12 shows the production volume of aggregate for concrete in Japan. In the 1960s, river gravel and sand were mainly used, but as they became depleted due to the construction boom during rapid economic growth, the main materials have been shifted today to crushed stone and sand. Sea gravel and sand have also been used in great quantities, but their extraction is likely to be totally banned in the near future as it has resulted in exposing bedrock on the seafloor, disappearance of sandbars, erosion of embankments, and damage to fishing grounds, resulting in a critical worsening of the marine environment.

Further, based on aggregate consumption in the United States since 1900 and its projection through 2020, as shown in Figure 3.13, the United States also saw a shift of the main materials from river gravel and sand to crushed stone and sand. The quantity of limestone and aggregate used for annual concrete production was calculated based on Figure 3.10, assuming that 1.1 tonne of limestone is required for production of 1 tonne of Portland cement, and that 0.3 and 1.9 tonnes of cement and aggregate are required, respectively, for the production of 1 m^2 of concrete.

TABLE 3.2

Four Scenarios of Global Socioeconomic Growth (IPPC, 2000)

Scenario A1: High-Growth Society

Economic growth continues and the world's population decreases after it reaches a peak in the mid-21st century.

New technology and high efficiency technologies are introduced rapidly.

Regional differences in per capita income are reduced by the reduction of local distance, the reinforcement of capacity, and the development of cultural and social exchanges.

Scenario B1: Sustainable Development-Type Society

The population growth and the rapid economic growth are the same as in the A1 scenario, but the substance-oriented factor is decreased.

Global measures are focused on the sustainability of economic, social, and environmental factors such as the introduction of clean and resource-saving technology.

Scenario A2: Social Diversification

An area is formed by blocks; each area maintains its own traditional culture and does not place a high value on economic efficiency which is based on free trade.

Energy dependence utilizes resources of the area.

Technological advances are minimal and based on population growth.

Scenario B2: Local Social Coexistence

Focus is on regional measures to achieve sustainability on economic, social, and environmental levels.

Population increases are modest in comparison with A2.

Economic development stops at intermediate level and is slower in comparison to A1 and B1; technological change is more extensive.

The quantities of limestone and aggregate used for concrete production worldwide per annum were approximately 1.65 and 9.5 billion tonnes, respectively, in 2000, but they are expected to rise in 2050 to approximately 6.0 and 16.2 billion tonnes, respectively (excluding use for roadbeds). As a large amount of aggregate is also used for non-concrete applications such as road construction, there is no data available regarding its global production quantity.

However, quantities cannot differ greatly between countries in terms of the percentage of materials required for social capital improvement. Thus, assuming that the ratio of cement production and aggregate consumption in the United States, Europe, and Japan (average of figures under column B/A in Table 3.1) can be applied "as is" to other countries, global aggregate production is estimated as approximately 45 billion tonnes. It is projected that, like cement consumption, the future production level of aggregate will also rise to approximately 58 billion and 84 billion tonnes in 2020 and 2050, respectively.

3.1.1.3 Energy Consumption

Energy issues correlate with all aspects of sustainability. Global warming is thought to be caused by CO_2 emissions generated from the consumption of

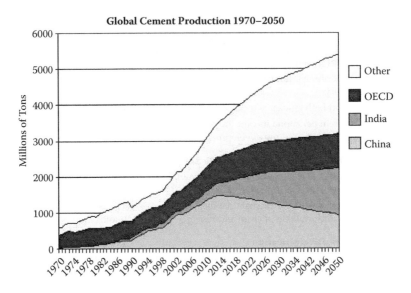

FIGURE 3.10
Global cement demand. (*Source:* Cement Sustainability Initiative, World Business Council for Sustainable Development. 2007. The Cement Sustainability Initiative. With permission.)

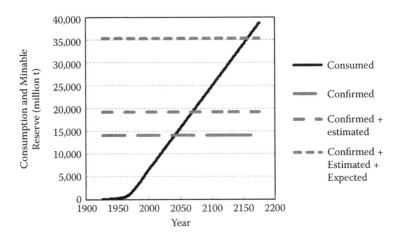

FIGURE 3.11
Predicted mineable reserves of limestone in Japan.

fossil energy (burning of fossil fuels). The issues also relate to resource recycling and waste problems; while energy sources such as fossil fuels and uranium are exhaustible resources, their extraction generates waste. As shown in Figure 3.14, with respect to the reserve-to-production ratio of energy resources (quantity of confirmed reserves divided by present consumption),

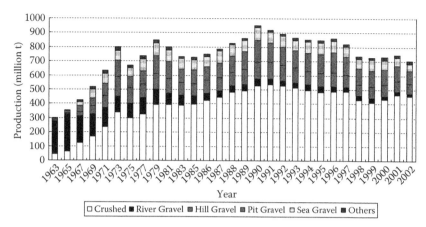

FIGURE 3.12

Japan aggregate production. (*Source:* Japan Crushed Stone Association. 2007. Changes in demand and supply of aggregate. http://www.saiseki.or.jp/kotsujukyu.html. With permission.)

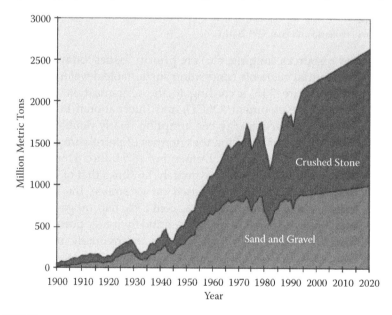

FIGURE 3.13

United States aggregate production. (*Source:* Tepordei, V. V. 1999. Natural Aggregates – Foundation of America's Future. U.S. Geological Survey Fact Sheet 144-97. With permission.)

the reserves are 41 years for petroleum, 192 years for coal, and 67 and 85 years for natural gas and uranium, respectively.

Furthermore, the supply of energy will have to be expanded due to growing demands for food in line with an increasing global population. The impact on biodiversity and changes in the ecosystem as a consequence of global warming resulting from the securing of energy supply sources (mining of

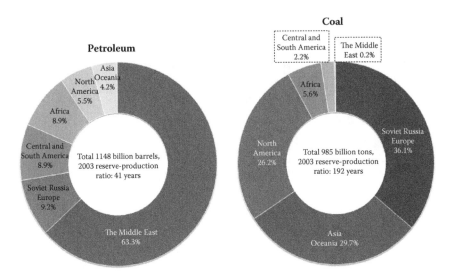

FIGURE 3.14
Reserves of petroleum and coal. (BP, 2004.)

underground resources, logging, etc.) are priority issues today. Thus energy is one of the essential elements concerning sustainable development.

As shown in Figure 3.15, according to the Organisation for Economic Co-operation and Development (OECD) and International Energy Agency (IEA), the world's primary energy consumption nearly doubled during the past thirty years, from 5.536 billion toes (tonnes of petroleum equivalent) in 1971 to 10.345 billion ktoes in 2002. Demand of 16.5 billion ktoes is expected in 2030, which is equivalent to approximately 1.6 times that of 2002.

Considering energy consumption based on its source, the 2002 ratios as shown in Figure 3.16 are: petroleum 36%, coal 23%, natural gas 21%, nuclear power 7%, and renewables including biomass and hydraulic power 13%. Of this renewable energy, sources other than biomass and hydraulic power (mainly solar and wind power) account for only 0.5% of total consumption. No great changes are foreseen concerning this composition ratio in the 2030 projection.

Taking the growing world population and expanding affluent lifestyles during the next fifty years into consideration, it is expected that the energy crisis will worsen markedly. Thus, in addition to the securing of supply sources, it will become indispensable for the demand sector to take measures also, and further promotion of energy conservation, utilization of nonfossil energy, and development of innovative energy technology are anticipated.

Figure 3.17 shows a breakdown of consumption based on energy source and demand sector. Energy consumption directly related to concrete structures falls under the "Industrial" and 'Transportation" sectors in the figure. The industrial sector includes the use of coal, petroleum coke, and heavy oil for clinker burning and independent power generators at cement plants;

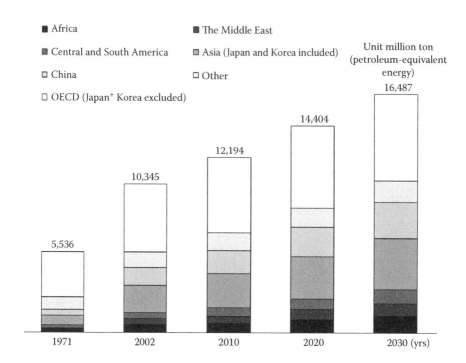

FIGURE 3.15
Energy consumption trends. (International Energy Agency, 2004.)

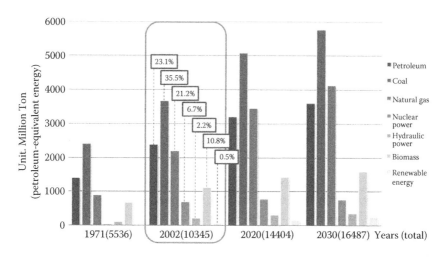

FIGURE 3.16
Energy consumption based on source. (International Energy Agency, 2004.)

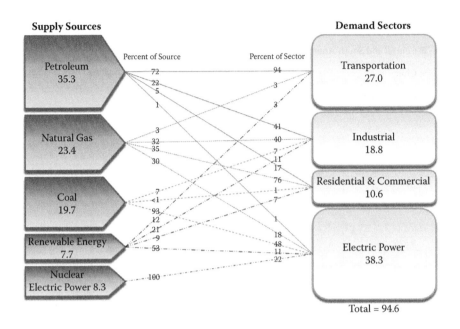

Supply Sources   **Demand Sectors**

FIGURE 3.17
Energy consumption and demand sector based on source.

gas oil for crushing machines and other heavy equipment at quarries and crushed stone and sand plants; coal, petroleum coke, and heavy oil for the manufacture of pig iron and steel, molding processes and independent power generators at reinforcing bar plants; kerosene for drying and heat compression processing of single panels at formwork plywood plants; and gas oil for construction machinery used for building concrete structures.

The transportation sector includes the use of gas oil and petrol for transporting cement and concrete. For example, in Japan, approximately 95 kg, 13 kg and 1.5 L of coal, petroleum coke, and heavy oil, respectively, and 30 kWh of purchased electricity are used to produce one tonne of Portland cement (Japan Cement Association, 2007). In crushed stone and sand plants, approximately 0.84 L, 0.21 L, and 0.03 L of gas oil, heavy oil, and kerosene and 2.5 kWh of purchased electricity are used to produce one tonne of aggregates (Japan Concrete Association, 2010).

Furthermore, 4.0 L of gas oil are consumed to transport 1 m³ of ready-mixed concrete (average return distance = approximately 21 km), and 4.5 kWh of purchased electricity are used to produce 1 m³ of ready-mixed concrete. Approximately 2.5 L and 0.3 L of gas oil and kerosene and 15 kWh of purchased electricity per floor area of 1 m² are used for construction machinery to build reinforced concrete buildings, while for the machinery to demolish the same area, approximately 2.8 L of gas oil per floor area of 1 m² are used (Japan Concrete Association, 2010). However, compared with these direct energy consumptions, amounts of energy consumed by air-conditioning, lighting, and hot-water supply during the use of concrete buildings are far greater. Such energy is covered by the residential and commercial sector and also by purchased electricity. Thus, a significant amount of energy classified in the electric power sector is eventually consumed during the use of concrete buildings.

3.1.1.4 Waste Generation

The reason waste problems attract public concern is because people, while hoping to achieve a recycling-oriented society, which plays a key factor in the creation of a sustainable society, acknowledge its difficulty in reality and also because they have concerns about environmental destruction caused by illegal dumping and adverse impacts resulting from the treatment and disposal of wastes typified by dioxins.

Such concern has generated a sense of avoidance concerning waste itself and waste treatment facilities (including intermediate processing facilities, incineration facilities, and final disposal sites). Especially when a waste generation point is far away from its incineration facility and final disposal site, the sense of burden bearing and unfairness among inhabitants of the surrounding area increases, hindering the establishment of a new facility or site. As revealed in Figure 3.18, the remaining capacity of final disposal sites has been decreasing every year, causing a shortage of disposal capacity and generating illegal dumping, resulting in a vicious cycle.

Gases generated from waste incinerating facilities include sulfur oxides (SOx), nitrogen oxides (NOx), hydrogen chloride, soot and dust (containing high levels of heavy metals), and dioxins. The emission of these substances is regulated by law. Another thing is that the construction of such disposal sites obliges changes in the natural environment through land use, while causing contamination of groundwater, bad odor, and methane gas generated by landfill garbage, dispersion of incinerator fly ash containing toxic

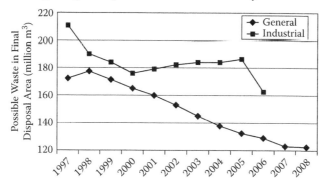

FIGURE 3.18
Decreasing capacity of disposal sites in Japan. (Ministry of the Environment, 2010.)

substances, and air and noise pollution due to concentration of waste haul-
ing vehicles. Thus, the treatment and disposal of waste usually causes various environmental impacts such as global warming as well as pollution and
contamination of the atmosphere, water, and soil.

One cannot make a realistic comparison, as statistical methods and classifications of waste types are different. However, 3 billion tonnes of solid
waste were generated in Europe in 2006 (6 tonnes per capita), 0.59 billion
tonnes in Japan in 2007 (5 tonnes per capita), and over 13 billion tonnes
annually (no accurate figure available) in the United States (49 tonnes per
capita) based on the Environmental and Energy Study Institute, 1995.
With recent global economic growth and population increases, mainly
in Asia, the amount of waste has augmented worldwide. According to
the forecast concerning non-industrial waste generated by OECD member
nations (OECD, 2008), the total amount of solid waste generated in 2005
was 1.7 times that of 1980, and is expected to increase 2.2 times in 2025. In
the same way, as shown in Figure 3.19, waste generated on a global level is
anticipated to increase from 12.7 billion tonnes in 2000 to 27 billion tonnes
in 2050.

Figure 3.20 shows the waste generation by type of activity in Europe,
revealing that the construction sector generates the most, accounting for
33% of the total. Similarly, Figure 3.21 shows the industrial waste generation by type of activity in Japan. Waste generated by the construction sector
accounts for 18.4% of total industrial waste and 13.1% of total waste, demonstrating that the construction sector is also a vital player in the reduction
of industrial waste generation. Breakdowns of the construction waste of

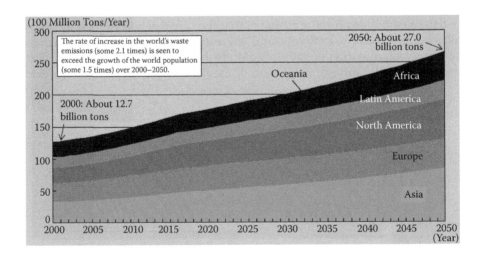

FIGURE 3.19
Projected global waste generation. (*Source:* Tanaka, M. 2011. Press release "A study on estimation and prediction of the amount of waste generated in the world." With permission.)

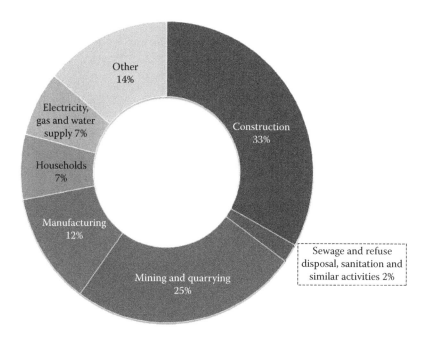

FIGURE 3.20
European waste generation categorized by activity. (*Source:* Eurostat, European Union. 2010. Environmental statistics and accounts in Europe. With permission.)

Europe in recent years and Japan since 1995 are shown in Figures 3.22 and 3.23, respectively.

Not apparent in Figure 3.22 are the specific features of countries, namely the Netherlands, Slovakia, Slovenia, and Bulgaria, that generate a significant percentage of construction sludge resulting from maintenance of their ports, rivers, and canals. As some demolished concrete rubble is reused at demolition sites, it is assumed that overall concrete waste generated from the demolition of buildings and structures amounts to two or three times the statistically obtained figure. After all, the fact that a large percentage of construction waste consists of mineral waste is universal at all times, and concrete and asphalt concrete account for 70 to 80% of the construction waste in many countries.

3.1.2 Environmental Impact

3.1.2.1 Greenhouse Gas Emissions

Infrared rays emitted into space from the Earth's surface are partially absorbed by greenhouse gases and raise temperature. Although the most representative greenhouse gas is carbon dioxide (CO_2), there are other gases with higher greenhouse effects such as methane gas (CH_4) with a warming

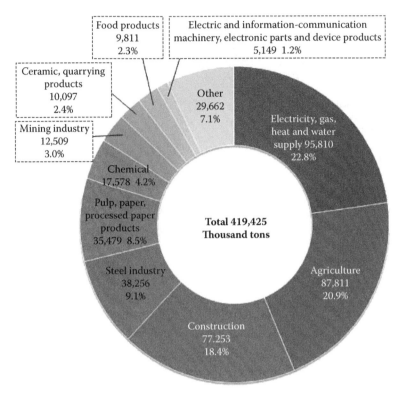

FIGURE 3.21
Japanese waste generation categorized by activity. (Ministry of the Environment, 2010.)

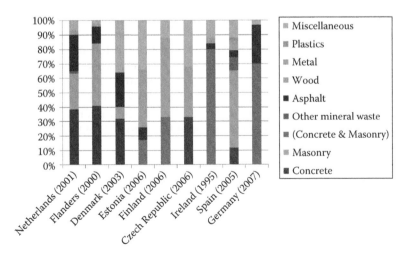

FIGURE 3.22
European construction waste by type. (European Commission, DG ENV, 2011.)

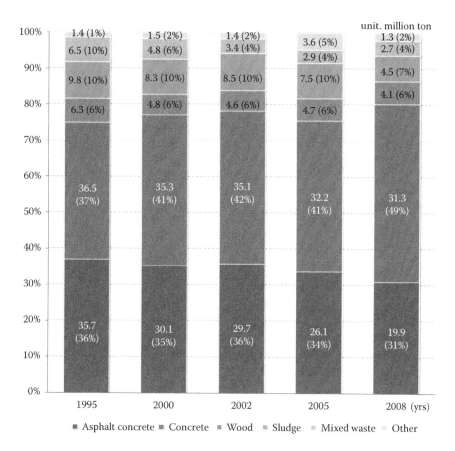

FIGURE 3.23
Japanese construction waste by type. (Ministry of the Environment, 2010.)

effect 21 times that of CO_2; nitrous oxide (N_2O), 210 times that of CO_2; and halocarbons (representative examples are chlorofluorocarbons), several hundred to several thousand times that of CO_2.

From the modern to contemporary era, the human-induced emission of these greenhouse gases has drastically increased due to the growing consumption of fossil fuels resulting from rapid economic and industrial development generated by intensifying human activities. Approximately 75% of human-induced CO_2 emissions during the past 20 years have been generated by the burning of fossil fuels. The increasing curves in CO_2 concentration during the past 50 and 2,000 years are shown in Figure 3.24. Based on this data, it is assumed that temperature increases, as shown in Figure 3.25, other climate changes, and incidental adverse effects have been seen in various areas, for example, submerging regions resulting from rising sea levels caused by retreating glaciers and melting of the permafrost, collapse of ecosystems, and more frequent droughts.

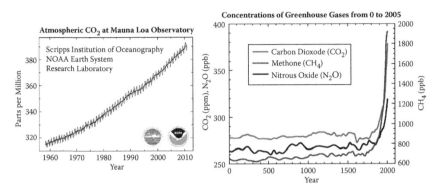

FIGURE 3.24

Carbon dioxide emission trends for 2,000 years. (*Source:* Left: Earth System Research Laboratoy, Global Monitoring Division, National Oceanic & Atmospheric Administration. 2011. http://www.esrl.noaa.gov/gmd/ccgg/trends/. With permission. Right: IPCC. 2007. Fourth Assessment Report (AR4), Climate Change 2007, The Physical Science Basis, Working Group I Contribution to the Fourth Assessment Report of the Intergovernmenal Panel on Climate Change. Cambridge University Press. With permission.)

The concentration of greenhouse gases in the atmosphere continues to increase at a soaring rate (CO_2 has been increasing at a rate of 1.5ppm/year during the past 20 years), and CO_2 concentration has increased from the 1750 level by 31%. Today's CO_2 level has reached a record high since 420,000 years ago, and it is highly likely that it is the highest since 20 million years ago.

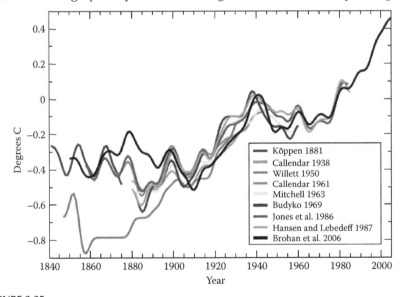

FIGURE 3.25

Temperature increase trends. (*Source:* IPCC. 2007. Fourth Assessment Report (AR4), Climate Change 2007, The Physical Science Basis, Working Group I Contribution to the Fourth Assessment Report of the Intergovernmental Panel on Climate Change. Cambridge University Press. With permission.)

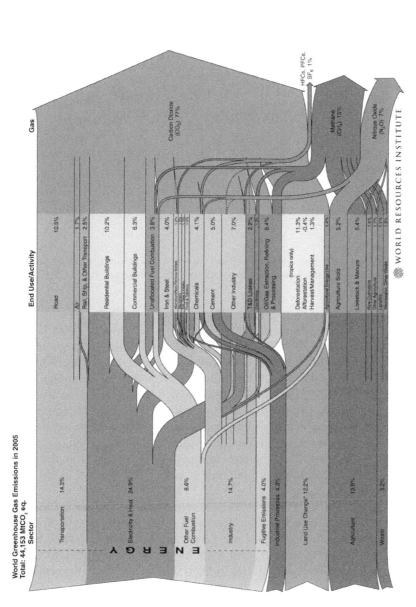

FIGURE 3.26
Global greenhouse gas emissions. (*Source:* World Resources Institute. 2005. Climate Analysis Indicators Tool, http://www.wri.org/chart/world-green-house-gas-emissions-2005. With permission.)

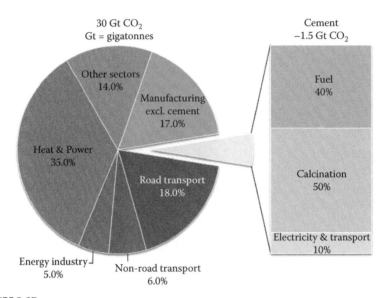

FIGURE 3.27
Carbon dioxide emissions from cement products. (*Source:* Cement Sustainability Initiative, World Business Council for Sustainable Development. 2005. Progress Report 2005. With permission.)

As shown in Figure 3.26, global greenhouse gas emissions from cement manufacturers were approximately 4% (5% as CO_2) of the total in 2000. Taking into consideration the increase in its production (from 1.6 billion tonnes in 2000 to 3 billion tonnes in 2010), the emissions are expected to be 7 to 8%—double the 2000 level. The breakdown of CO_2 emissions from cement manufacturers, as shown in Figure 3.27, consists of those generated by the combustion of fossil fuels during cement production (approximately 40%), those due to generation of electricity required for cement production accounting (approximately 5%), and those from the combustion of fuels during transportation of cement materials (approximately 5%). The remaining 50% is generated from the reaction of calcination of the key raw material of cement, limestone ($CaCO_3$). The relevant chemical equation is $CaCO_3 \rightarrow CaO + CO_2$.

In Japan, as shown in Figure 3.28, CO_2 emissions generated in relation to buildings account for more than one-third of the total amount. The figure reveals that the emissions generated by energy consumption needed to heat, cool, and light buildings during their life spans are particularly high, accounting for two-thirds of building-related CO_2 emissions. This situation is not limited to Japan.

3.1.2.2 Destruction of Ozone Layer

In the stratosphere where strong ultraviolet rays from the Sun are present, an oxygen molecule dissociates into two oxygen atoms by absorbing ultraviolet rays, and these atoms then bond with other oxygen molecules to produce

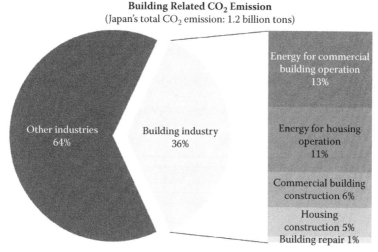

FIGURE 3.28
Carbon dioxide emissions from occupied buildings in Japan. (Architectural Institute of Japan, 2008.)

ozone. In the ozone layer located at an altitude of approximately 15 to 50 km in the stratosphere, an ozone molecule dissociates into one oxygen molecule and one oxygen atom by absorbing ultraviolet rays. In this way, ozone protects life on Earth through both its production and dissociation by absorbing ultraviolet rays.

Chlorofluorocarbons (CFCs) are artificial substances with a high chemical stability. Because they are inexpensive and harmless to humans, their production rapidly increased from 1960, as shown in Figure 3.29, and they are used for a broad range of applications, including refrigerants for refrigerators

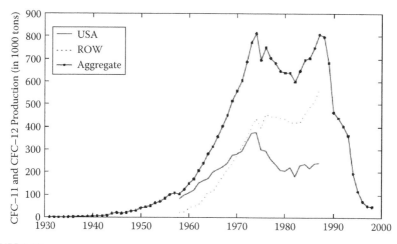

FIGURE 3.29
Chlorofluorocarbon (CFC) production since 1960. (Auffhammer, M. et al., 2005.)

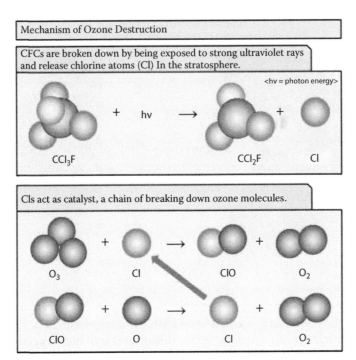

FIGURE 3.30
Chlorofluorocarbon (CFC) and halon emissions.

and air-conditioners, blowing agents for building insulators, and propellants for spray cans. However, because of their stable property, CFCs ascend into the stratosphere when released after use.

Bromofluorocarbons (halons) are colorless, odorless, and mildly toxic substances with extremely strong fire-extinguishing properties. Like CFCs, they reach the stratosphere by global scale atmospheric circulation with little dissociation in the troposphere. As shown in Figure 3.30, CFCs and halons reaching the stratosphere are broken down by exposure to strong ultraviolet rays and release chlorine atoms (Cl) and bromine atoms (Br). These Cls and Brs act as catalysts, successively breaking down ozone molecules, causing a decrease of ozone in the stratosphere and ultimately depleting the ozone layer.

Ultraviolet (UV) light has a wavelength (approximately 100 to 400 nm) shorter than that of visible light, and the shorter its wavelength, the higher its energy becomes, which affects living organisms. UV-C light which has a short wavelength (100 to 290 nm) is completely absorbed by oxygen molecules in the ozone layer and atmosphere and does not reach the Earth. UV-A light with a long wavelength (320 to 400 nm) can cause creases and sagging of the skin, but as it is hardly absorbed by the ozone layer, decreasing ozone in the stratosphere has very little affect on its intensity on the Earth's surface. On the other hand, the intensity of UV-B light (wavelength of 290 to 320 nm) increases under the influence of ozone layer depletion, causing damage to

DNA, inhibition of photosynthesis, metabolic abnormalities, and induction of apoptosis that may be responsible for sunburn, skin blemishes, freckles, skin cancer, cataracts, and declines in the function of the immune system.

It is said that when the ozone layer decreases by 1%, the amount of UV-B reaching the Earth increases by 1.5%. In addition, UV rays affect plastic materials, causing deterioration, decoloration and denaturation, resulting in lower strength and shorter life spans.

The Vienna Convention for the Protection of the Ozone Layer reached an agreement in 1985 for the purpose of designating substances believed to be responsible for ozone depletion, and controlling their manufacture, consumption, and trading. Based on this, the Montreal Protocol on Substances That Deplete the Ozone Layer was adopted in 1987. Following a phased tightening of regulations and revisions, requests were made to abolish totally the use of specified CFCs, halons, and carbon tetrachloride by 1996 among developed countries (by 2015 among developing countries), and HCFCs, alternative materials for CFCs, by 2020 among developed countries (by 2030 in principle among developing countries). However, as revealed in Figure 3.31, depletion of the ozone layer is still progressing today, and in Figure 3.32, it is reported that its concentration decreased by 50% over the Arctic Circle between March 2010 and March 2011. It is vital to further control the release of ozone-depleting materials.

Ozone layer depletion problems related to concrete include the releases of CFCs used as refrigerants for old type air-conditioners installed in concrete

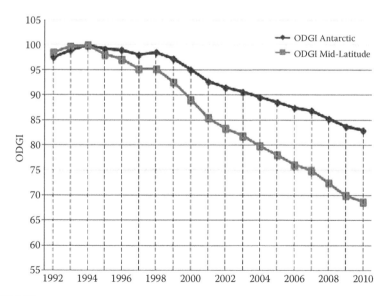

FIGURE 3.31
Ozone layer depletion. (*Source:* Earth System Research Laboratory, Global Monitoring Division, National Oceanic & Atmospheric Administration. 2011. http://www.esrl.noaa.gov/gmd/odgi/. With permission.)

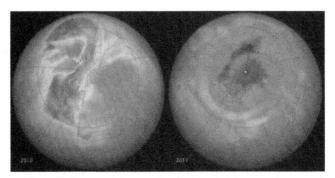

FIGURE 3.32
Arctic Circle ozone layer depletion, 2010–2011. (*Source:* NASA Earth Observatory. 2011. Arctic Ozone Loss, March 30, 2011, http://earthobservatory.nasa.gov/IOTD/view.php?id=49874. With permission.)

buildings; releases of CFCs from foam insulation during demolition of concrete buildings; and releases of the halon fire extinguishing agent used for concrete building fires.

In response to the Montreal Protocol, various countermeasures have been implemented, for example, recovery and destruction of fluorocarbons from freezers and air-conditioners during demolition and renovation of buildings; appropriate management, recovery, and reuse of halons in extinguishing equipment; and utilization of recycled halons. At the same time, development and use of hydrofluorocarbons (HFCs) have been promoted as replacements for CFCs and halons. However, as HFCs act as strong greenhouse gases, the use of materials other than fluorocarbons has also been promoted for air-conditioner refrigerant and insulation blowing agents. Currently, fire-extinguishing agents equivalent to halons in terms of performance and safety have not yet been developed.

3.1.2.3 Acidification (Acid Rain)

Acid rain is a phenomenon in which emissions of acidic substances originating from sulfur oxides (SOx) and nitrogen oxides (NOx) released into the atmosphere through the burning of fossil fuels and refining of metals are dissolved in rain, snow, and mist. As a result, it acidifies rivers, lakes, and soil while directly affecting buildings and structures including cultural assets. The impact of acid rain extends to regions of emerging economies such as China and Southeast Asian countries, in addition to advanced industrial countries in Europe and North America. As its causative substances SOX and NOx can be carried several thousand kilometers, it is a global concern that transcends national borders.

When falling on rivers and lakes, acid rain causes the decrease or death of aquatic species (see Figure 3.33), while producing genetic defects and increases in abnormalities typified by hermaphroditism. In the case of

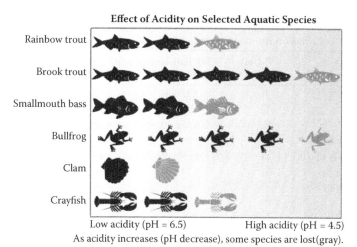

Low acidity (pH = 6.5)   High acidity (pH = 4.5)
As acidity increases (pH decrease), some species are lost(gray).

FIGURE 3.33

Acidification impacts on aquatic species. (*Source:* U.S. Geological Survey. 1998. Resources, FY 1998 Annual Financial Report, http://pubs.usgs.gov/98financial/resources. html. With permission.)

forests and woods, it kills trees and contaminates the soil, causing desertification and resulting eventually in the decrease and death of wildlife. In soil, nutrients react with acid and run off, causing damage such as growth inhibition and decrease in crop yields. When it permeates the underground water, it contaminates drinking water, increasing risk to human health. On the sea, it causes massive outbreaks of toxic plankton, resulting in the decrease and death of marine life. Acid rain can also seriously damage and deteriorate buildings and artifacts made of concrete or marble, as well as copper roofs and aluminium construction materials.

With respect to concrete, SOx and NOx are emitted when burning fuels, such as coal, gas oil, heavy oil, kerosene, and natural gas for the operation of cement and concrete manufacturing equipment, motor vehicles, and ships for their transportation, as well as machinery for the construction and demolition of concrete structures. SOx and NOx are also emitted when incinerating waste generated in relation to these activities. Furthermore, we should not forget that SOx and NOx emissions are caused by thermal power generation, when purchased electricity is used for the production of cement and concrete, and building of concrete structures.

Table 3.3 shows examples of the SOx and NOx emission intensity per unit in Japan (Japan Society for Civil Engineers, 2004), generated as a result of the production of various materials, operation of concrete manufacturing and construction machinery, and transportation vehicles and ships. In Japan, approximately 6% of total domestic NOx emissions result from the operation of construction machinery. In the case of a 100 m^3 reinforced concrete structure, as shown in Table 3.4 (Japan Society for Civil Engineers, 2004), the

TABLE 3.3

SOx and NOx Emission Intensity in Japan

Product or Activity	Unit (x)	SO$_x$ Emission (kg SOx/(x))	NO$_x$ Emission (kg NOx/(x))	PM Emission (kg PM/(x))
Portland cement	T	1.22E-01	1.55E-00	3.58E-02
Fly ash[a]	T	6.20E-03	7.54E-03	1.25E-03
Crush stone	T	6.07E-03	4.15E-03	1.41E-03
Reinforcing bar	T	1.34E-01	1.24E-01	1.01E-02
Concrete mixer (3.0 m³)	m³	1.99E-04	2.44E-04	4.58E-05
Concrete pumping vehicle (piping 90 to 100 m³/h)	m³	2.54E-04	4.94E-03	2.50E-04
Crawler crane (rope system, 16 t hanging)	Hour	1.37E-02	2.67E-01	1.35E-02
Backhoe excavator (0.6 m³)	Hour	3.98E-02	7.74E-01	3.93E-02
Diesel truck (10 t)	Km/t	9.41E-05	9.14E-04	7.85E-05
Concrete mixer truck (4.5 m³)	Km/m³	1.95E-04	3.79E-03	1.92E-04
Motor vehicle and ship (10,000 t)	Km/t	3.44E-04	5.77E-04	8.86E-06

[a] As fly ash is a by-product of a thermal power plant, SOx, NOx and PM produced by thermal power generation are not considered.

TABLE 3.4

SOx, NOx, and PM Emissions and Percentages from 100m³ Reinforced Concrete Building

	SO$_x$		PM		NO$_x$	
	Emission (kg)	Percentage	Percentage	Emission (kg)	Emission (kg)	Percentage
Material production	7.66	63.2	40.6	49.85	1.58	27.9
Building construction	1.16	9.5	16.0	19.61	1.10	19.2
Building demolition	1.81	14.9	28.7	35.25	1.79	31.4
Material disposal and recycle	0.23	1.9	1.6	2.00	0.12	2.1
Transport of materials, concrete, waste	1.27	10.5	13.1	16.11	1.10	19.4
Total	10.87			106.72	4.59	

NOx emissions from the construction and demolition of reinforced concrete buildings account for 50% of their entire life cycles. On the other hand, the SOx emissions from the production of materials such as cement and reinforcing bars account for two-thirds.

3.1.2.4 Outdoor Air Pollution

Air pollution is an environmental problem that has worsened with expanding urban and industrial activities, particularly with growing fossil fuel consumption. The phenomenon involves contamination of air with SOx, NOx, and suspended particulate matter (SPM) of 10 μm or less contained in car exhaust and industrial smoke, adversely affecting human health and the living environment, as well as animals and plants. During the mid-twentieth century, a large number of people suffered illness and death due to severe air pollution throughout the world. Especially well known is the Great Smog of London in December 1952. A massive amount of sulfur dioxide and soot emitted from coal burning by industry and for domestic heating and cooking formed a thick layer of smog to the extent that it reduced visibility and killed about 4,000 people, mainly children and the elderly.

In the latter half of the 1950s, as energy sources shifted from coal to petroleum, many petrochemical complexes were built, where SOx was emitted due to the burning of sulfur-rich oil, damaging human health by asthma, chronic bronchitis, and heart diseases. NOx and volatile organic compounds (VOCs) from car exhaust and industry produced smog consisting of a photochemical oxidant (major component: O_3), causing human suffering including irritation of the eyes, nose, and throat, and asthma, chronic bronchitis, decline in lung function, as well as the inhibition of plant growth.

In addition to the aforementioned SOx and NOx emissions, examples of particulate matter emissions are also shown in Tables 3.3 and 3.4. In recent years, asbestos used for insulation and fireproofing in the past has been dispersed into the air from construction materials during the renovation and demolition of buildings, posing another problem.

3.1.2.5 Water Pollution and Soil Contamination

Water pollution occurs when the biological purification system provided by plankton living in the sea and rivers no longer works and the water quality deteriorates, through contamination caused by toxic substances contained in domestic and industrial wastewater. It also occurs when the soil is contaminated by hazardous substances (e.g., dioxine, PCB, lead, and arsenic) that then permeate the underground water.

On the other hand, eutrophication, a type of water pollution, occurs when the flow of nutrient salts (mainly nitrogen and phosphorus) increases into enclosed water areas such as inland bays and lakes, raising their concentration. When eutrophication progresses, it interferes with the normal material cycle of the aquatic ecosystem while increasing the phytoplank-ton and organic constituents present in water, causing red tides that trigger decreases in the level of dissolved oxygen, eventually changing the natural ecology. The sources of pollutants contributing to eutrophication include discharges from homes, industry, livestock, agriculture, effluents from cit-ies, and acid rain.

Soil contamination is caused by volatile organic compounds (VOCs), heavy metals (including lead, hexavalent chrome, and mercury) and oil contained in industrial effluent and waste. The characteristics of soil contamination are (1) it is not likely to travel or spread easily, (2) it accumulates in the soil over a long period, and (3) it exerts a greater impact on private assets (it potentially lowers asset value), compared with air and water pollution. In some cases, it affects underground water and causes severe damage to human health. Thus, soil contamination has a negative impact on the human body, land asset value, and company valuation.

In ready-mixed concrete plants, wastewater is generated when washing concrete mixers and transporter vehicles. On construction sites of concrete structures, sludge water results from the high-pressure washing of concrete joints, concrete cutting by waterjet machines, and the washing of concrete pumping vehicles and piping. All of the waste fluid is alkaline; it passes through a neutralization process before being discharged. On the other hand, during the building of some concrete structures, concrete piles are installed by the earth drilling method using a stabilizer; foundation improvement work is implemented using a cement-based soil solidifier; and earth retain-ing and waterproofing work for underground areas is carried out.

In such cases, the concern arises that heavy metals (chrome is most criti-cal) contained in cement or concrete may seep into the underground water upon contact. Also, at final waste disposal sites or ground where crushed concrete from demolition is used for roadbeds, attention should be given to the seeping of heavy metals into the soil. Such releases of heavy metals can occur not only with ready-mixed but also with hardened concrete, and can be carried by underground water and also rainwater. The types of heavy metals and their quantities contained in cement, aggregate, and fly ash are shown in Table 3.5 (fib, 2003). The amount of chrome released from ready-mixed concrete varies, depending on the conditions where it is poured, as shown in Table 3.6 (fib, 2003).

3.1.3 Disruption of Ecosystem (Biodiversity)

Living organisms evolved with the Earth over approximately 4 billion years and differentiated in various ways, resulting in the formation of life. The phenomenon of life is in fact based on three different diversity levels; genetic, species, and ecosystem. All these function in totality, although they interact very closely and intricately.

TABLE 3.5

Heavy Metals Contained in Cement and Aggregate Fly Ash (mg/kg)

Heavy Metal	Cement Type			
	Portland	Portland Blast Furnace	Silica-Based Aggregate	Fly Ash
Sb	2–18	–	0.16–2	<0.5–37
As	4–23	0.8–2	0.5–11	0.6–321
Ba	185–390	–	3–121	612–2249
Cd	0.03–6	<0.1–1	0.01–1	0.2–7
Cr	25–124	20–80	0.7–70	29–360
Co	2.8–21	<0.2–2	–	36–125
Cu	55–98	5–17	0.6–39	38–613
Ga	5–9	–	–	2–84
Pb	5.0–254	<1–18	0.7–50	46–301
Hg	<0.02–0.12	<0.1	<0.01–0.1	<0.5–0.7
Ni	17–97	4–25	0.7–50	46–301
Tl	<0.2–4.1	<0.5	<0.1–1	0.7–4
Zn	21–679	5–80	1–60	47–1483

As biodiversity not only maintains ecological balance, but also influences the climate system and provides mankind with natural blessings, it is an important aspect of sustainability. Therefore, conservation of the three diversities and maintaining them in a sound condition means maintaining nature in a beautiful and healthy condition and also securing the resources biodiversity brings to human society and the fundamental natural infrastructure, while enabling continuous supplies of a multitude of ecosystem (or environmental) services. Indeed, almost all human activities, including the economy, are based on biodiversity, the basis of life.

Human socioeconomic activity, development, and land utilization (modification of the land surface or its retention in a non-natural condition) induce the destruction or loss of nature. They cause the extinction of flora and fauna and their habitats and biotopes, resulting in decreases and extinctions of species of plants and animals, destruction, fragmentation, and deterioration of ecosystems, and other indirect damages caused by changes in weather and insolation.

TABLE 3.6

Chrome Emissions from Surfaces of Ready-Mixed Concrete (mg/m²)

Concreting Measure	Without Retarder (24 hours)	Long-Term Retarded Concrete (72 hours)
Against soil with defined water content	–	≤1.3
Against water-saturated soil	≤11.1	≤17
Against flowing groundwater	30–50	–

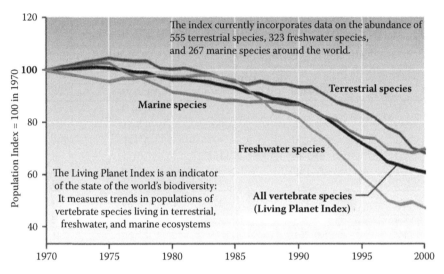

FIGURE 3.34
Effects of anthropogenic activities on biodiversity. (*Source:* World Resources Institute. 2005. Millennium Ecosystem Assessment, Ecosystems and Human Well-being: Biodiversity Synthesis. With permission.)

Deforestation results from the use of timber for building materials and processed goods, clearance for quarrying, and the expansion of urban areas. This reportedly represents 50% of the factors responsible for the extinction of species of plants and animals. Deforestation also changes landscapes and academically and historically valuable geological formations and conditions. Advancement of the urbanization of forests and sloping land causes landslides, while the development of river zone farmland for housing lowers the retaining and permeating functions of rainwater and causes urban flooding.

As a result of these human activities, biodiversity has been gradually collapsing, and as shown in Figure 3.34, in 2000, the total number of vertebrae species decreased to 60% of the species existing in 1970.

It is necessary to acknowledge that the extraction of materials for production of cement and concrete and construction of concrete structures directly and indirectly cause collapses of biodiversity. The direct causes include mining of limestone for the main ingredient of cement, and rock for crushed aggregate, which in most cases involves deforestation and destruction of the ecosystem; the extraction of sand and gravel used as aggregate from rivers and marine areas and the construction of dams and river embankments which reduce habitats in river areas; the construction of transportation networks such as roads and railways, resulting in the fragmentation of ecosystems; and urban development that changes the utilization of land in the surrounding areas and has a negative impact on wildlife (see Figure 3.35).

The indirect causes include the global warming resulting from CO_2 emissions generated by cement production and transportation of concrete

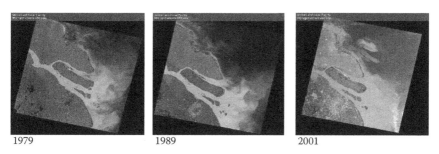

FIGURE 3.35
Changes in land use by urban development around Shanghai. (LANDSAT WRS-1, MSS by USGS acquired from ESDI at Global Land Cover Facility, University of Maryland.)

materials and the destruction of ecosystems caused by air pollution and acid rain due to SOx and NOx emissions.

3.1.4 Other Environmental Impacts

3.1.4.1 Heat Island

An urban heat island is, as shown in Figure 3.36, a phenomenon in which the temperature of an inner urban area becomes higher than the suburbs, arising from a thermal imbalance of its ground surface. Heat island occurs more significantly during the night than in the daytime and when there is little wind.

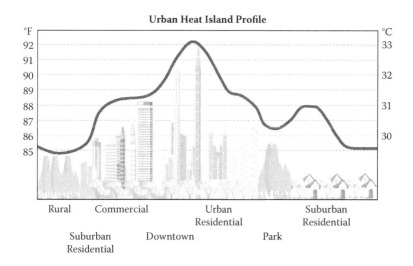

FIGURE 3.36
Heat island.

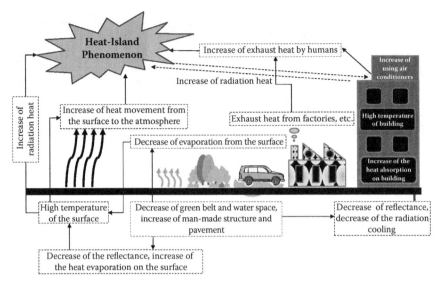

FIGURE 3.37
Causes of heat island phenomenon.

The causes are, as shown in Figure 3.37, an increase in anthropogenic heat emissions by air conditioners (approximately one-third of solar radiation heat); the accumulation of solar radiation heat due to covering of ground surfaces by asphalt and concrete (the heat storage capacity of concrete is 2,000 times that of air); solar radiation heat repeatedly reflected on the surfaces of high-rise buildings, then captured and absorbed by the buildings; deterioration of ventilation caused by high-rise buildings; and less temperature reduction by water evaporation due to decreased green areas.

Acceleration of the urban heat island phenomenon generates a change in the wind direction pattern of the perimeter area and an increase in the probability of rain in the urban center while causing stress or heat stroke in humans and changes in the ecosystem.

3.1.4.2 Noise and Vibration

Noise, meaning unpleasant sound, is defined by frequency (in kilohertz [kHz] units) and intensity (sound pressure). Usually, sound intensity is measured by the sound pressure level (in decibel [dB] units) expressed on a logarithmic scale to adapt to the human acoustic sense. As shown in Table 3.7, 0 dB is the threshold of the sound pressure level of human hearing, and it is said that humans feel pain when sound exceeds the level of 125 dB. The noise tolerance level in various living conditions and lifestyles varies, depending on the time of day, location environment, duration of noise, and other factors. Table 3.8 shows the tolerance level responding to location environment set by the Swiss Noise Abatement Ordinance, (Swiss Federal Council, 1986).

TABLE 3.7

Examples of Sound Pressure Levels

Sound Pressure (Pa)	Sound Pressure Level (dB)	Sound Source
	170	
1000 Pa	160	Shot gun
	150	Pistol
100 Pa	140	Bolting gun
	130	Jet engine test stand
10 Pa	120	Pain limit
	110	Pneumatic drill jumbo
1 Pa	100	Chain saw fuel powered
	90	Wood cutting machine tool
100 mPa	80	Milling machine
	70	Road traffic
10 mPa	60	Conversation
	50	Office
1 mPa	40	Living room
	30	Reading room
100 µPa	20	Sleeping room
	10	Radio studio
20 µPa	0	Hearing threshold

Vibration on the other hand refers to tremors of the ground and structures sensed by humans. It is said that humans sense vibration in the frequency range of 0.1 to 500 Hz, and the levels that produce problems are in the range of 1 to 90 Hz. Vibration can be unpleasant; it causes more psychological and sensory problems than physiological ones. It induces problems in daily life such as emotional disorders including irritation, difficulty in falling asleep, easily waking, and interference with thinking and working. It also causes property damage to furniture and other household items and cracks in concrete and finishing materials.

The sources of vibration, often the same as those of noise, are typically industrial plants, construction sites, and traffic as shown in Table 3.9. Unlike

TABLE 3.8

Noise Tolerance Levels

Sensitivity Level	Planning Value in dB (A)		Noise Impact Threshold in dB (A)		Alarm Level in dB (A)	
	Day	Night	Day	Night	Day	Night
I (recreational zones)	50	40	55	45	65	60
II (residential zones)	55	45	60	50	70	65
III (mixed residential industrial zones)	60	50	65	55	70	65
IV (industrial zones)	65	55	70	60	75	70

TABLE 3.9

Main Vibration Sources

Source	Examples
Factory	Air blower, compressor, pressing machine, crushing machine, attritor, kneading mixer, cutting machine
Construction site	Pile driving, pile extracting, sheeting, ground compaction, blasting, demolition
Transportation	Large automobile, rail system

noise, vibration usually propagates in the ground, generally within a range 100 m from its source and in most cases it propagates between 10 and 20m. However, aircraft can transmit vibrations through air as the medium without propagating through the ground and cause buildings to vibrate. Vibration also varies in other propagation characteristics, depending on the geological structure and on the differences of the natural vibration frequency between the ground and building. Types of tremors also depend on a building's structure (reinforced concrete, steel-frame, masonry, or wood). Finding countermeasures is made more challenging by such variable vibration factors.

Vibration and noise related to concrete and concrete structures include those generated when cement and aggregate are crushed and classified in the course of their production process and their loading and unloading; when using construction machinery for drilling, foundation piling, and concreting during construction of concrete structures (Figure 3.38); by vehicles and trains, i.e., during the service life of concrete structures such as road and railway bridges; and when using construction machinery for the demolition of concrete structures.

3.1.4.3 Landscape Destruction (Aesthetic Degradation of Landscape)

Placing priority on economics rather than the pursuit of a good landscape and environment, has caused the destruction of the natural landscape by the haphazard extraction and mining of soil and stones; destruction of urban, rural, and natural landscape by the construction of buildings and civil engineering structures including houses, high-rise buildings, industrial plants, revetments, roads, and bridges, without due consideration concerning their harmony with the region, aesthetic quality, or tradition; and destruction of the natural landscape by the haphazard designation of final waste disposal sites.

Certain cultural and natural assets registered as UNESCO World Heritage sites in accordance with the Convention Concerning the Protection of the World Cultural and Natural Heritage are deemed to have significant and universal value and should be preserved and passed on to future generations. Sites that are critically endangered because of large-scale construction work and urban development are placed on the World Heritage in Danger list (see Table 3.10).

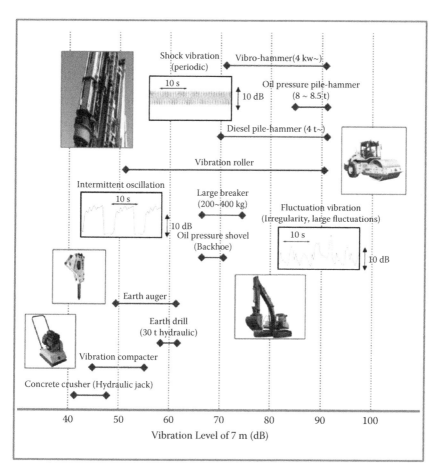

FIGURE 3.38
Vibration and noise impacts.

Cologne Cathedral (Figure 3.39), registered as a World Heritage site in 1996, was added to the list of the World Heritage in Danger in 2004, as it was threatened by a plan to build high-rise buildings that would have destroyed the surrounding landscape. However, following the serious efforts exerted by the city authorities who imposed a height regulation for surrounding buildings, the cathedral was removed from the danger list in 2006.

The historic town of Zabid in Yemen has the great mosque Al Asa'ir, known as the first university in the Arabian Peninsula, and its traditional houses have exterior walls in typical Zabid style, laid with small bricks and plastered over with stucco with geometric designs. However as some of these walls have been replaced with concrete, the town was added to the list of the World Heritage in Danger in 2000. The Dresden Elbe Valley, registered as a World Cultural Heritage site in 2004, was deleted from the

TABLE 3.10

World Heritage in Danger Sites Affected by Construction and Urban Development

Country	Site	Listing Date and Status	Reason for Listing
Bulgaria	Srebarna Nature Reserve	1992–2003 (released)	Construction of bank; ground pollution from agrochemicals
Ecuador	Sangay National Park	1992–2005 (released)	Road construction
Tunisia	Ichkeul National Park	1996–2006 (released)	Construction of dam; urbanization
Ethiopia	Simien National Park	1996 (current)	Poaching; war; expansion of agricultural land; urban development; population growth
Brazil	Iguazu National Park	1999–2001 (released)	Planned road construction
India	Monuments at Hampi	1999–2006 (released)	Construction of cable suspension bridge; road construction; changes in agricultural land; destruction of natural features
Democratic Republic of the Congo	Salonga National Park	1999 (current)	Poaching; urbanization (residential construction)
Yemen	Historic Town of Zabid	2000 (current)	Urbanization; deterioration; increased concrete construction
Pakistan	Fort and Shalimar Gardens in Lahore	2000 (current)	Aging of fort; urban development; fountain damage at Shalimar caused by road expansion
Algeria	Tipasa	2002–2006 (released)	Lack of effective management plan; inadequate maintenance; vandalism and defacement; urban development
Afghanistan	Minaret and Archaeological Remains of Jam	2002 (current)	Damage and digging caused by continuing war; river flooding; planned road through archaeological area
Nepal	Kathmandu Valley	2003–2007 (released)	Construction without due consideration of harmony with region
Azerbaijan	Walled City of Baku	2003–2009 (released)	Damage from large earthquake; urban development; lack of protective policy
Iraq	Ashur (Qal'at Sherqat)	2003 (current)	Risk of submersion by construction of large dam
Germany	Cologne Cathedral	2004–2006 (released)	Loss of city landscape by construction of high-rise buildings
Venezuela	Coro and its Port	2005 (current)	Damage from heavy rains; development of surrounding area

(continued)

TABLE 3.10 (CONTINUED)

World Heritage in Danger Sites Affected by Construction and Urban Development

Country	Site	Listing Date and Status	Reason for Listing
Germany	Dresden Elbe Valley	2006–2009 (deleted)	Bridge construction
Senegal	Niokolo-Koba National Park	2007 (current)	Potential for environmental deterioration from construction of Gambia River dam
Georgia	Bagrati Cathedral and Gelati Monastery	2010 (current)	Reconstruction plan that may damage authenticity and completeness

list in 2009, because the construction of a bridge across the valley to reduce the traffic congestion in the city of Dresden was considered to spoil the integrity of the landscape (Figure 3.40).

3.1.4.4 Indoor Air Pollution

In order to enhance the efficiency of indoor heating and cooling by preventing heat intake or release, the airtightness of buildings has been greatly improved. As a result however, the indoor–outdoor ventilation air volume was reduced while the indoor air pollution level remained, resulting in sick building syndrome. The pollutants consist of chemical and biological substances. Building materials (mainly adhesives, paints, and floor materials) and furniture that contain large amounts of volatile organic compounds

FIGURE 3.39

Cologne Cathedral, Germany, now on UNESCO's World Heritage in Danger list.

FIGURE 3.40
Elbe River bridge near Dresden, Germany.

(VOCs) such as formaldehyde and toluene can cause vertigo, nausea, and headache, and may possibly elevate the incidence of cancer. Table 3.11 shows indoor VOC concentration level guidelines in Japan.

Fungi and bacteria such as molds and microorganisms may also be responsible for allergies and infectious diseases.

In addition, long-term or repeated exposure to air containing radon (radioactive gas) generated from radium that may be found in stones and tiles and large amounts of asbestos fibers released from refractory covering materials and boards during renovation or demolition can cause lung cancer.

3.2 Social and Economic Aspects

Concrete allows the creation of free-form structures and is moreover robust and durable in character. These two basic qualities have brought marked increases in concrete consumption. Civil engineering structures are used for extended periods as the basis for human social and economic activities. Buildings are constructed as places for the manufacture of industrial products, the management of business activities, the educational and cultural activities, the generation and dissemination of information, the daily living, and other purposes, all of which are essential to humans in contemporary society. The appropriate functioning of infrastructure and buildings makes human social and economic activity possible. Conversely, when buildings and infrastructure are not available, the resulting social and economic losses are immense.

TABLE 3.11

Indoor VOC Concentration Level Guidelines in Japan

Volatile Organic Compound	Major Building Material Uses and Sources	Indoor Concentration Level Guidelines
Formaldehyde (HCHO)	Plywood, particle board, decorated plywood, laminated lumber, insulator (glass fiber), complex flooring, starch adhesive for wallpaper	$100\ \mu g/m^3$ (0.08 ppm)
Toluene ($C_6H_5CH_3$)	Timber preservatives, oil varnishes, adhesion bonds	$260\ \mu g/m^3$ (0.07 ppm)
Xylene ($C_6H_4(CH_3)_2$)	Timber preservatives, oil paints and varnishes, plastic paints, adhesion bonds	$870\ \mu g/m^3$ (0.20 ppm)
Paradichlorobenzene	Tick and insect repellents, air fresheners	$240\ \mu g/m^3$ (0.04 ppm)
Ethyl benzene ($C_6H_5C_2H_5$)	Organic solvents, adhesion bonds	$3800\ \mu g/m^3$ (0.88 ppm)
Styrene (C_6H_5-CH=CH_2)	Expanded polystyrene, raw material of synthetic rubber	$220\ \mu g/m^3$ (0.05 ppm)
Chlorpyrifos	Ant repellents (organic phosphorus)	$1\ \mu g/m^3$ (0.07 ppb)
		$0.1\ \mu g/m^3$ (0.007 ppb) : in the case of children
Di-n-butyl phthalate		$220\ \mu g/m^3$ (0.02 ppm)
Tetradecane ($C_{14}H_{30}$)		$330\ \mu g/m^3$ (0.04 ppm)
Di-2-ethylhexyl phthalate		$120\ \mu g/m^3$ (7.6 ppb)
Diazinon		$0.29\ \mu g/m^3$ (0.02 ppb)
Acetaldehyde (CH_3CHO)		$48\ \mu g/m^3$ (0.03 ppm)
Fenobucarb		$33\ \mu g/m^3$ (3.8 ppb)

The construction of infrastructure and buildings also creates jobs; many people are engaged in the relevant planning, design, construction work, and maintenance and management. The construction field generally provides a large amount of employment, and in Japan, represents approximately 8% (five million employees) of total employment. This is approximately half of the 16% accounted for by the manufacturing industry, thus representing a very high proportion. There are no relevant statistics, but given that concrete is used in the construction of almost all infrastructure and buildings, the amount of employment directly and indirectly related to concrete accounts for a considerable proportion.

In United Kingdom, 2.2 million people are engaged in the construction industry, which is an important component of the economy (Department for Business, Innovation and Skills, 2010). United Kingdom's construction industry also represents around 8% or a little less of total employment, and thus is similar to that in Japan. Generally, the construction industry's share in employment differs with the economic and industrial structure of a country; however, it is one of the prime industries and has a very great influence on society and the economy.

The safe and reliable nature of concrete makes possible the creation of convenient and pleasant urban spaces and living environments. Appropriate design that takes into account the external forces (such as seismic forces) acting on concrete structures ensures structural safety, in other words, human safety. The safety of a structure is the most important element of its function, and the concrete and construction sector has drawn on a wide range of experiences to formulate design standards for concrete structures. Even today, structures with insufficient seismic resistance are sometimes destroyed by earthquakes. However through seismic retrofitting, the seismic resistance of such structures has steadily improved, providing also protection against typhoons and hurricanes, as well as excellent fire-resistance, thereby assuring a sense of security.

The development of high-rise concrete buildings limits uncontrolled use of land and makes its efficient use possible. Making cities more compact increases the efficiency of working environments and as a result reduces environmental burden. Improved sewer provision significantly improves the urban sanitary environment and greatly reduces the incidence of infectious disease. Meanwhile, the flood prevention and irrigation made possible by the construction of dams secures safe living and working environments and makes agricultural production possible. This generates enormous economic benefit. When disaster prevents the utilization of infrastructure and buildings, all everyday social and economic activities are affected. In this way, concrete structures and the networks they support are indispensable to social life.

Social aspects also include cultural factors. Everyone recognizes the cultural value of infrastructure and buildings built in the Roman era. There are many examples where the cultural value of such heritage is protected by retaining only the façade as a cultural relic and inserting a concrete structure in the interior.

Meanwhile, concrete is also used to prevent the spread of hazardous materials. For instance, the containment vessels of nuclear power stations are required to have excellent load-bearing ability in order to block the spread of radioactivity, and are generally made of reinforced concrete with steel liners. Currently, nuclear power generation is the focus of attention as a way of generating electricity without producing CO_2, and concrete plays a pivotal role in terms of safety, not only in containment vessels but throughout nuclear power facilities. Safe and stable electricity generation by nuclear power stations minimizes economic costs across the board. However, if that safety is ever lost, the resulting social impact and economic cost are immeasurable.

Safety inevitably involves externalities not envisaged at the time of construction, and which may in the future impose a very great social and economic burden. This has been graphically illustrated by the accident at the Fukushima nuclear power station in the wake of the Great East Japan Earthquake, which struck on March 11, 2011.

References

Architectural Institute of Japan. 2008. Recommendations for Environmentally Conscious Practice of Reinforced Concrete Buildings.

Auffhammer, M., Morzuch, B.J. and Stranlund, J.K. 2005. Production of Chlorofluorocarbons in Anticipation of the Montreal Protocol. *Environmental and Resource Economics*, Vol.30, No.4, pp.377-391.

BP. 2004. Energy Outlook 2030.

CEMBUREAU. 2009. Activity Report 2009.

Cement Sustainability Initiative, World Business Council for Sustainable Development. 2005. Progress Report 2005.

Cement Sustainability Initiative, World Business Council for Sustainable Development. 2007. The Cement Sustainability Initiative.

Cohen, J.L. and Moeller, G.M. 2006. *Liquid Stone, New Architecture in Concrete*. Princeton Architectural Press. New York.

Committee on Resource Utilization Strategy, Architectural Institute of Japan. 2008. Challenge of Material Flow and Resource Utilization in Building Materials.

Department for Business, Innovation, and Skills, United Kingdom. 2010. *Low Carbon Construction: Final Report.*

Earth System Research Laboratory, Global Monitoring Division, National Oceanic & Atmospheric Administration. 2011. http://www.esrl.noaa.gov/gmd/ccgg/trends/.

Earth System Research Laboratory, Global Monitoring Division, National Oceanic & Atmospheric Administration. 2011. http://www.esrl.noaa.gov/gmd/odgi/.

Environmental and Energy Study Institute. 1995. 1995 Briefing Book on Environmental and Energy Legislation.

ERMCO. 2002. European Ready-mixed Concrete Industry Statistics Year 2001.

European Commission, DG ENV. 2011. Final Report Task 2 - Service Contract on Management of Construction and Demolition Waste - SR1.

Eurostat, European Union. 2010. Environmental statistics and accounts in Europe.

fib. 2003. Bulletin 23, Environmental Effects of Concrete.

International Energy Agency. 2004. World Energy Outlook 2004.

IPCC. 2000. Summary for Policymakers, Emissions Scenarios, A Special Report of Working Group III.

IPCC. 2007. Fourth Assessment Report (AR4), Climate Change 2007, The Physical Science Basis, Working Group I Contribution to the Fourth Assessment Report of the Intergovernmental Panel on Climate Change. Cambridge University Press.

Japan Cement Association. 2007. Summary of Cement LCI Data.

Japan Concrete Institute. 2010. Committee Report on Minimization of Global Warming Substances and Wastes in Concrete Sector.

Japan Crushed Stone Association. 2007. Changes in demand and supply of aggregate. http://www.saiseki.or.jp/kotsujukyu.html.

Japan Society for Civil Engineers. 2004. Assessment for Environmental Impact of Concrete, Concrete Engineering Series 62.

Kelly, T. 1998. Crushed Cement Concrete Substitution for Construction Aggregates–A Materials Flow Analysis, U.S. Geological Survey Circular 1177.

Matos, G.R. 2009. Use of Minerals and Materials in the United States from 1900 through 2006. U.S. Geological Survey Fact Sheet 2009-3008.

Ministry of the Environment, Government of Japan. 2009. Establishing a sound material-cycle society, Creating economic development through the establishment of a sound material-cycle society.

Ministry of the Environment, Government of Japan. 2010. Annual Report on the Environment, the Sound Material–Cycle Society and the Biodiversity in Japan, Our Responsibility and Commitment to Preserve the Earth–Challenge 25.

Ministry of the Environment, Government of Japan. 2010. Establishing a Sound Material-cycle Society, Milestone Toward a Sound Material–Cycle Society through Changes in Business and Life Styles.

NASA Earth Observatory. 2011. Arctic Ozone Loss, March 30, 2011, http://earthobservatory.nasa.gov/IOTD/view.php?id=49874.

OECD. 2008. OECD Environmental Outlook to 2030.

Swiss Federal Council. 1986. Noise Abatement Ordinance 814.41.

Tanaka, M. 2011. Press release "A study on estimation and prediction of the amount of waste generated in the world".

Tepordei, V.V. 1997. Natural Aggregates–Foundation of America's Future. U.S. Geological Survey Fact Sheet 144-97.

U.S. Energy Information Administration. 2010. Annual Energy Review 2009.

U.S. Geological Survey. 1998. Resources, FY 1998 Annual Financial Report, http://pubs.usgs.gov/98financial/resources.html.

U.S. Geological Survey. 2002. *Minerals Yearbook*, Vol.1, Metals & Minerals, Cement.

World Resources Institute. 2005. Climate Analysis Indicators Tool, http://www.wri.org/chart/world-greenhouse-gas-emissions-2005.

World Resources Institute. 2005. Millennium Ecosystem Assessment, Ecosystems and Human Well-being: Biodiversity Synthesis.

4

Evaluation Systems of Sustainability

4.1 Life Cycle Assessment (LCA)

Since the Industrial Revolution, mankind has established a social system based on mass production, mass consumption, and mass disposal. However, the increasing global population and rising living standards have led to a vast increase in resource and energy consumption. As a result, global environmental problems have become major issues for mankind, so much so that, if the growth in resource and energy utilization continues unabated, it is feared that these resources will become exhausted and that global pollution and global warming will become serious problems. To make sustainable development possible going forward, mankind needs as far as possible to reduce inputs from nature and outputs to nature in all activities. This requires that we measure input and output in all our activities, assess their impacts on the global environment, and take action to reduce them. It was against this background that life cycle assessment (LCA) came into being.

Under the International Organization for Standardization (ISO) 14040 standards (1997), LCA is defined as "compilation and evaluation of the inputs, outputs and the potential environmental impacts of a product system throughout its life cycle." Put another way, LCA is a form of technology for assessing the environmental aspects of and associated latent burden of the object concerned, and its scope generally extends to extraction of resources, production of materials, manufacture of products, use, end of life phase, and disposal. It has recently become common practice to include in this approach, known as the "cradle-to-grave" concept, the extent of recycling and re-use also, in which case the LCA would be from "cradle-to-cradle." Figure 4.1 presents a flow chart of these processes. The 1997 ISO standards list the following four advantages of carrying out LCA:

1. Identifying opportunities to improve the environmental aspects of products at various points in their life cycles
2. Decision making in industry, governmental or non-governmental organizations

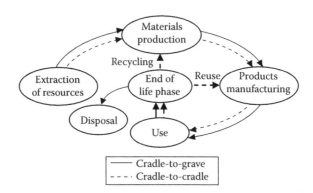

FIGURE 4.1
Life cycle flow chart.

3. Selection of relevant indicators of environmental performance including measurement technique

4. Marketing

LCA consists of the following three stages:

1. Compiling an inventory of relevant inputs and outputs of a product system

2. Evaluating the potential environmental impacts associated with inputs and outputs

3. Interpreting the results of inventory analysis and impact assessment phases in relation to the objectives of the study

These processes can be represented in the framework shown in Figure 4.2. The first task is to clarify the ultimate goal of carrying out the LCA and the scope of the assessment. When doing so, the functions and functional units of the subject system must be specified. Then to establish system boundaries, a decision is made as to what unit processes are to be included in the LCA,

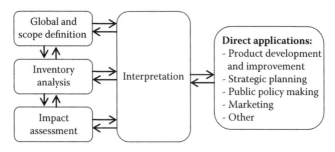

FIGURE 4.2
Life cycle assessment processes.

and a life cycle inventory analysis is carried out. Life cycle inventory analysis is defined as the "phase of life cycle assessment involving the compilation and quantification of inputs and outputs, for a given product system throughout its life cycle."

The next task is the life cycle impact assessment, which is the phase of LCA aimed at understanding and evaluating the magnitude and significance of the potential environmental impacts of a product system. This process includes assigning inventory data to impact categories and applying weighting factors so as to achieve an overall assessment. Lastly, the findings of the inventory analysis and impact assessment are brought together to carry out an interpretation. The results are presented in the form of conclusions or recommendations. All LCA processes are basically iterative and the content they examine can be modified as required. Naturally, an appropriate review is required to assess whether the results of LCA are valid.

The following items may be among the objects of environmental impact assessment to be considered under LCA:

Global climate change

Natural resources use (materials, water, fuels)

Stratospheric ozone level

Land use and habitat alteration

Eutrophication

Acidification

Air pollution

    Smog (tropospheric ozone formation)

    Particulate matter air pollution

Other air pollution (toxics and other compounds)

    Indoor air pollution

Water pollution

Soil contamination

Pollution due to radioactive substances

Impacts due to waste generation

Noise and vibration

4.2 Environmental Standards for Buildings

The standard for general products and services stated in the previous section was developed by ISO/TC207. It is therefore not easy to apply it directly to the construction sector. To resolve this difficulty, ISO/TC59 (building construction)/SC14 (design life) developed an environmental standard for buildings (ISO 2004) that sets an assessment method for the potential environmental impacts of constructed assets at the construction stage. Figure 4.3 illustrates the overall framework for technical, economic, and environmental assessment in service life planning.

The basis of this approach is to examine whether the performance levels of the designed constructed asset satisfy all requisite performances as determined by the requirements based on the client briefs, regulations, and targets. In practice, this will usually mean studying a number of options and choosing the most suitable. This means that the functional equivalency of the constructed asset under consideration is important. Requirements relating to environmental impact may be expressed under headings such as those below.

Use of materials

Use of energy

Use of water

Emissions including hazardous and toxic substances

Use of land and impact on biodiversity

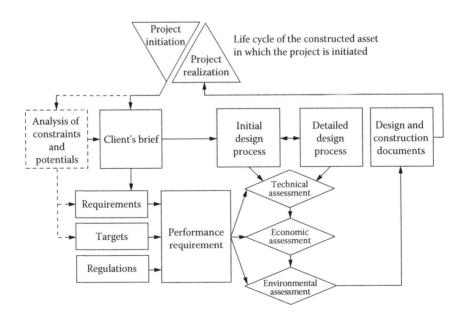

FIGURE 4.3
Framework for service life planning.

The approach to LCA is the same as that outlined in Section 4.1, but given the particular features of constructed assets, the environmental impacts will normally arise from the following activities:

Raw material extraction

Material production

Construction

Use and maintenance

End of life

To assess environmental impacts at each stage, the inventory data required for the LCA must be supplied. It is important for the inventory data to be objective and transparent, but even within the same activity, there is generally variation in the data. In some cases, past data can be used but in other cases new data must be obtained. The appropriate data must be used in accordance with the particular features of the object to be assessed and the requirements.

Meanwhile, ISO/TC59/SC17 (sustainability in buildings and civil engineering works) developed a standard on the environmental declaration of building products (ISO 2007). This standard establishes principles and requirements concerning type III environmental declarations (EPDs) of building products. The term "product" used alone relates not only to goods or product systems but can also include service systems. This standard introduced the idea of a product category representing a group of building products that can fulfill equivalent functions. In order to carry out an actual LCA, product category rules for the following items must be laid down:

Product category definition and description

Goal and scope definition for LCA of product

Inventory analysis

Others

The EPD must include the following environmental information:

Environmental impacts, for example, climate change

Use of resources and renewable primary energy

Waste to disposal

Emissions to water, soil, and indoor air

Additional environmental information

The development of type III environmental declarations is voluntary.

4.3 Systems of Environmental Impact Evaluation

Reducing environmental impact requires a system for the practical application of environmental impact assessments. In 1998, the ISO issued standard ISO 14020 (ISO 2000) on environmental labels and declarations designed to provide information on the overall environmental character, a specific environmental aspect, or any number of aspects of a product or service. This standard is basically designed for application to a single industrial product and service, and not for complex items such as civil engineering structures and buildings. The environmental evaluation tools for buildings shown in Section 4.6 can be seen as an evolved form of this standard.

ISO 14020 specifies the purpose of environmental labels and declarations as follows: "The overall goal of environmental labels and declarations is, through communication of verifiable and accurate information, that is not misleading, on environmental aspects of products and services, to encourage the demand for and supply of those products and services that cause less stress on the environment, thereby stimulating the potential for market-driven continuous environmental improvement." Nine principles are set out to ensure that this purpose is reliably met. Labels and declarations must:

1. Be accurate, verifiable, relevant, and not misleading
2. Not be subject to procedures or requirements that create unnecessary obstacles to international trade
3. Be based on scientific methodology
4. Make all information available
5. Consider all relevant aspects of the life cycle of the product
6. Not inhibit innovation
7. Not be precluded by administrative requirements or information demands
8. Include open participatory consultation with interested parties in development processes
9. Make available information on the environmental aspects of products and services to purchasers and potential purchasers

Environmental labels and declarations fall into the following three types:

Type I environmental labeling (ISO 1999)

Type II environmental labeling (ISO 1999)

Type III environmental declarations (ISO 2007)

Type I environmental labeling is a voluntary, multi-criteria-based third party program that authorizes display of the eco-mark (Japan), green seal (U.S.), EU eco-label, etc.

Type II environmental labeling consists of self-declared environmental claims that include statements, symbols, and graphics relating to the product. In general, the terms used in the claim may include phrases such as "extended life product," "recovered energy," "recyclable," "recycled content," "recycled material," "reduced energy consumption," "reduced resource use," "reduced water consumption," "reusable," and "waste reduction," all of which have specific requirements. However, as sustainability is still at the research stage, the standard states that claims relating to sustainability are not permitted. Times have clearly changed. The standard stipulates that the Möbius loop (symbol in the shape of three twisted chasing arrows forming a triangle) may be used only in relation to claims on recycled or recyclable content. Under Type II environmental labeling, different countries use various labeling symbols.

Type III environmental declarations indicate the environmental burden of a product by giving a quantitative assessment of environmental impact based on ISO 14040 series standards. In other words, they indicate life cycle environmental impact information (raw material extraction, production, use, disposal, and recycling). They are designed mainly for use in business-to-business communication, but use in business-to-consumer communication is not precluded under certain conditions. In Japan, the Ecoleaf program (JEMAI, 2011) is an example of this type of declaration.

This kind of labeling reveals its value only after actual use has begun. In other words, by making preferential use of labeled products, general consumers and government bodies can effect reductions in environmental burdens, which as a result gives producers incentives to further reduce environmental burdens. As price is generally a strong priority in product choice for consumers, the system does not always function effectively. Nevertheless, use of labeled products can be noted by an enterprise in its CSR report as publicity for its stance on the environment, thus promoting green procurement of products. Meanwhile, government bodies engaged in public works or similar projects practice green procurement as an environmental strategy. For example, concrete blocks made with waste material are often used as green procurement products. In the past, however, environmental impact was not always quantitatively assessed, so that such products were used without an exact idea of the extent of environmental impact reduction. Recently, the importance of quantitative assessment in green procurement has come to be recognized, and a shift in that direction has gathered pace.

4.4 Environmental Standards for Concrete Sector

Since 2008, ISO/TC71/SC8 (environmental management for concrete and concrete structures) has been engaged in the development of an environmental standard for the concrete sector. This move arose from the decision to establish rules "for the concrete sector by the concrete sector" as the ISO 13315 series. This decision was taken in light of the view that it was not easy to carry out an environmental impact assessment for the concrete sector based on the ISO 14000 series or allow people in the concrete sector to carry out such assessments appropriately and with proper understanding. The ISO 13315 series consists of the following parts.

Part 1: General principles

Part 2: System boundary and inventory data

Part 3: Constituents and concrete production

Part 4: Environmental design of concrete structures

Part 5: Execution of concrete structures

Part 6: Use of concrete structures

Part 7: End-of-life phase including recycling of concrete structures

Part 8: Labels and declaration

Figure 4.4 shows the basic framework of the ISO 13315 series. At present, Part 1 has already been published and Part 2 is under development. This standard series is, of course, being coordinated for consistency with the ISO 14000 series and other related standards.

Figure 4.5 presents a flowchart of environmental design under ISO 13315-1. This framework is consistent with the basic approach of ISO 15686-6. To understand the application of this method, we will present the example of a reinforced concrete rigid frame viaduct (fib, 2008). This is for the construction of a viaduct at a railway crossing (Figure 4.6a), and we will assume that the required environmental performance specified is a 20% reduction in CO_2 emissions compared to a conventional structure (Figure 4.6b). To satisfy this requirement, a structural type that ensures required seismic performance by increasing the bearing capacity of piles with a new technology and reinforcing the joints between piles and columns with steel tubes was introduced (Figure 4.6c). A calculation of CO_2 for the new structure indicated a 28% reduction compared to a conventional structure, as shown in Figure 4.6d.

In other words, the environmental performance (R) of this new structure satisfied the required environmental performance (S). In general, the setting of the required environmental performance is determined by regulation, the owner, the designer, etc. The verification in environmental design is sometimes

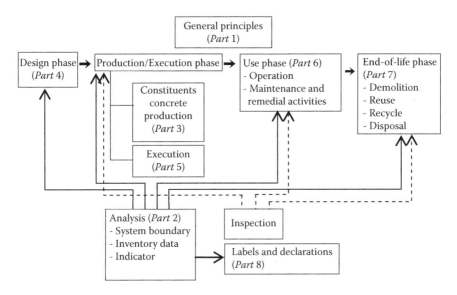

FIGURE 4.4
Basic framework of ISO 13315 series.

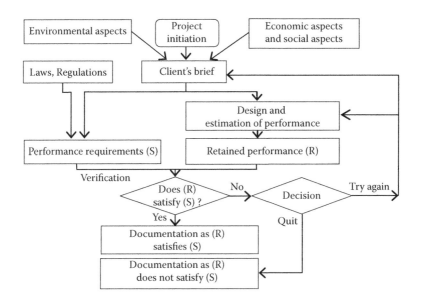

FIGURE 4.5
Environmental design flow chart for ISO 13315-1.

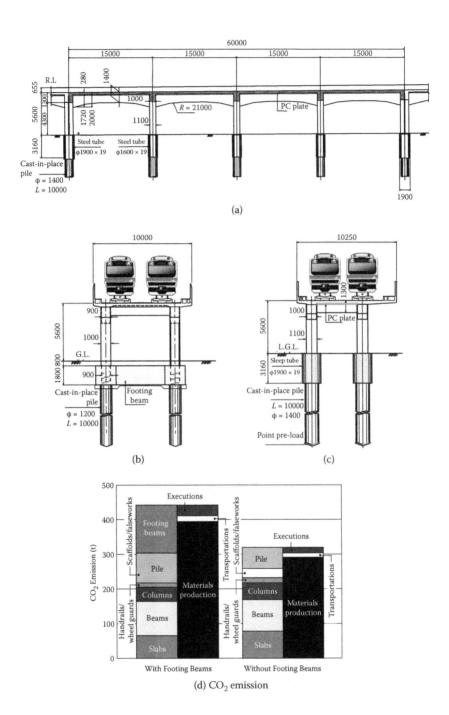

FIGURE 4.6
(a) Railway crossing viaduct. (b) Conventional structure with footing beam. (c) New structure without footing beam. (d) Projected carbon dioxide reductions in new structure.

not understood, but in the interest of environmental impact reduction, it is essential to set a target of some kind as the performance requirement.

4.5 Evaluation of Social and Economic Aspects

Assessment of these factors, i.e., the social and economic aspects that comprise sustainability, is also important in addition to the environmental aspect. There is a varied range of potential indicators for assessing social aspects. The following are examples of potential assessment items.

Quality of buildings as places to live and work

Safety and security

Indoor conditions

Serviceability

Accessibility

Access to services needed by users

Architectural quality

Cultural characteristics

Social cohesion and connectedness

Users' acceptance of infrastructures and satisfaction

Protection of cultural heritage

Others

The items listed above are all assessment indicators for application after completion of a constructed asset, but during the construction of the asset, the following indicators may be applied:

Safety of laborers

Inconvenience

Others

Indicators for assessment of economic aspects may include the following:

Performance

Location

Energy efficiency

Maintenance

Functionality

Others

Building performance greatly influences a building's life cycle cost, and the quality of the products and services as well as intellectual property created within the building. Building location influences land cost and also the costs of transporting people and goods. Energy efficiency greatly influences building operation cost. Similarly in the case of infrastructure, performance greatly influences life cycle cost. As a piece of infrastructure is generally used longer than a building, increasing the initial investment by a modest amount can reduce life cycle cost. Infrastructure location and functionality have a marked influence on the efficiency and cost of socio-economic activities. Appropriate maintenance of infrastructure can also minimize the deterioration and interruption of the services provided by the infrastructure, and thereby reduce economic cost.

Until now, the three factors comprising sustainability have been discussed separately, but they are as a matter of course interrelated. To assess sustainability, they need to be considered as a whole, but it is not so easy to do this in a rational manner.

4.6 Tools of Environmental Effect Evaluation for Buildings

Tools to assess the environmental impact of buildings have been developed and are in widespread use. As it is not possible to present them all here, an outline will be given of the representative assessment tools that are in frequent use at present: (1) the Building Research Establishment Environmental Assessment Method or BREEAM; (2) the Comprehensive Assessment System for Building Environmental Efficiency or CASBEE; and (3) Leadership in Energy and Environmental Design or LEED.

4.6.1 BREEAM

This is an environmental assessment tool for buildings developed by Britain's Building Research Establishment. According to the BREEAM official website (2011), this tool was developed for the purpose of providing clients, developers, designers, and others with the following:

Market recognition for low environmental impact buildings

Assurance that best environmental practice is incorporated into a building

Inspiration to find innovative solutions that minimize the environmental impact

A benchmark that exceeds regulation

A tool to help reduce running costs and improve working and living environments

A standard that demonstrates progress toward corporate and organizational environmental objectives

The objects of BREEAM assessment are as follows and can be applied in various situations.

Whole new buildings

Major refurbishments of existing buildings

New-build extensions to existing buildings

A combination of new-build and existing building refurbishment

New-builds or refurbishments that are parts of larger mixed use buildings

Existing building fit-out

The BREEAM assessment categories and main issues are as shown in Table 4.1. For each item, a score (%) is calculated based on achieved credits as a proportion of available credits with reference to a preset weighting. The score achieved is given a rating: unclassified (<30), pass (≥30), good (≥45), very good (≥55), excellent (≥70), and outstanding (≥85). The conditions vary with the object of the assessment.

4.6.2 CASBEE

CASBEE (2011) was developed by a committee at the Institute of Building Environment and Energy Conservation under the guidance of the Japanese Ministry of Land, Infrastructure, Transport, and Tourism. The basic tools are predesign (PD), new construction (NC), existing building (EB), and renovation (RN). The objects of the CASBEE basic assessment are energy efficiency, resource efficiency, outdoor environment, and indoor environment, but the method is not to adopt a direct assessment of these factors. Instead, the object of the assessment is divided into interior and exterior areas demarcated by "virtual boundaries" defined for instance by the site boundaries, and each of these areas is assessed from the following perspectives.

1. Building environmental quality and performance or Q (quality)
2. Building external environmental load or L (loadings)

Q indicates the level of living comfort for the building users and is assessed under three item headings: indoor environment (Q1), quality of service (Q2), and outdoor environment on site (Q3). L represents the negative aspects of the environmental impact on the exterior and is also assessed under three item headings: energy (L1), resources and materials (L2), and off-site environment (L3). There is a specified weighting coefficient for each item. Table 4.2

TABLE 4.1

BREEAM Categories and Main Issues

Management	**Waste**
Commissioning	Construction waste
Construction site impacts	Recycled aggregates
Building user guide	Recycling facilities
Health and Wellbeing	**Pollution**
Daylight	Refrigerant use and leakage
Occupant thermal comfort	Flood risk
Acoustics	NOx emissions
Indoor air and water quality	Water course pollution
Lighting	External light and noise pollution
Energy	**Land Use and Ecology**
CO_2 emissions	Site selection
Low or zero carbon technologies	Protection of ecological features
Energy submetering	Mitigation/enhancement of ecological value
Energy-efficient building systems	
Transport	**Materials**
Public transport network connectivity	Embodied life cycle impacts of materials
Pedestrian and cyclist facilities	Reuse
Access to amenities	Responsible sourcing
Travel plans and information	Robustness
Water	**Innovation**
Consumption	Exemplary performance levels
Leak detection	Use of BREEAM accredited professionals
Reuse and recycling	

shows the items evaluated under Q and L classes. Under CASBEE, building environmental efficiency (BEE) is assessed using BEE = Q/L.

Ideally a building should provide a high level of living comfort and impose a low impact on the exterior. The BEE value increases with an increase in the value of Q and a decrease in the value of L. As shown in Figure 4.7, environmental labeling under the BEE value is divided into the following categories: C (poor), B⁻ (fairly poor), B⁺ (good), A (very good), and S (excellent).

4.6.3 LEED

LEED is an energy and environmental design rating system developed by the U.S. Green Building Council (USGBC). According to the LEED official website (2011), its "green building certification program encourages and accelerates global adoption of sustainable green building and development

TABLE 4.2

CASBEE Assessment Items

Q (Quality)			L (Loadings)		
Q1	Q2	Q3	L1	L2	L3
Sonic environment	Serviceability Durability and reliability	Preservation and creation of biotope	Building thermal load	Water resources	Consideration of global warming
Thermal comfort	Flexibility and adaptability	Townscape and landscape	Natural energy utilization	Reducing use of non-renewable resources	Consideration of local environment
Lighting and illumination		Local characteristics and outdoor amenities	Efficiency in building service systems	Avoiding materials containing pollutants	Consideration of surrounding environment
Air quality			Efficient operation		

practices through a suite of rating systems that recognize projects that implement strategies for better environmental and health performance." The assessment objects of LEED for new buildings are as follows:

Homes

Neighborhood development (in pilot)

Commercial interiors

Core and shell

New construction

Schools, healthcare, retail

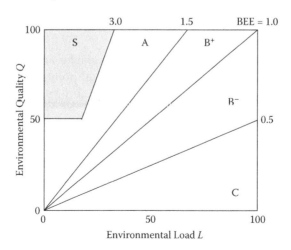

FIGURE 4.7

CASBEE environmental labeling.

There is also an assessment system relating to the operations and maintenance of existing buildings. For instance, according to the LEED 2009 New Construction and Major Renovation Checklist, the assessment items and respective numbers of points are as follows.

Sustainable sites (26)

Water efficiency (10)

Energy and atmosphere (35)

Materials and resources (14)

Indoor environmental quality (15)

Innovation in design (6)

Regional priority (4)

Of these, materials and resources receive the following assignments of items and points (maximum 14 points).

Credit 1.1 Building reuse – Existing walls, floors and roof (1–3)

Credit 1.2 Building reuse – Maintain existing interior nonstructural elements (1)

Credit 2 Construction waste management (1–2)

Credit 3 Materials reuse (1–2)

Credit 4 Recycled content (1–2)

Credit 5 Regional materials (1–2)

Credit 6 Rapidly renewable materials (1)

Credit 7 Certified wood (1)

Additionally, the storage and collection of recyclables are prerequisites. The following ratings are given according to the total points of each item: certified (40–49), silver (50–59), gold (60–79), and platinum (80 and above).

References

BREEAM. 2011. http://www.breeam.org/page.jsp?id=66

CASBEE. 2001. http://www.ibec.or.jp/CASBEE/english/index.htm

fib. 2008. Environmental design of concrete structures: general principles, Technical Report, Bulletin 47.

ISO. 1997. Environmental management. Life cycle assessment: principles and framework. ISO 14040.

ISO. 1999. Environmental labels and declarations. Type I environmental labeling: principles and procedures. ISO 14024.

ISO. 1999. Environmental labels and declarations. Type II environmental labeling: principles and procedures. ISO 14024.

ISO. 2000. Environmental labels and declarations: general principles. ISO 14020.

ISO. 2004. Buildings and constructed assets: service life planning. Part 6: Procedures for considering environmental impacts. ISO 15686-6.

ISO. 2006. Environmental labels and declarations. Type III environmental declarations: principles and procedures. ISO 14025.

ISO. 2007. Sustainability in building construction. Environmental declaration of building products. ISO 21930.

JEMAI. 2001. http://www.jemai.or.jp/english/ecoleaf/index.cfm

LEED. 2001. http://www.usgbc.org/DisplayPage.aspx?CMSPageID=222

5

Technologies for Concrete Sustainability

5.1 General

Concrete sustainability involves all activities in the processes of concrete raw material production, concrete production, and the construction, demolition, recycling, reuse, and disposal of its structures. The key is to satisfy requirements with as little consumption of resources and energy as possible.

Minimum consumption of resources and energy first brings environmental benefits; in other words, more resources will be conserved while reducing hazardous emissions such as CO_2, NOx, SOx, and dust into the atmosphere, water, and soil. Naturally this will produce great and direct economic advantages. The construction of concrete structures will then provide the infrastructure to support various social activities. Furthermore, once this support is indicated in visible ways, the concrete sector will enhance its social value as a sustainable industry that is capable of realizing sustainable development.

As concrete is the major material used in concrete structures, it has to meet requirements for specific performance regarding strength and durability. The performance of concrete differs greatly, depending on its ingredients and their mixture proportions, and can be evaluated based on four elements, i.e., strength, durability, fire-resistance, and environmental performance. It is necessary to determine the performance of concrete in light of the requirements concerning a given structure.

Regarding the performance of concrete as a material, the cement content usually has to be increased in order to obtain high strength, which enhances durability but decreases environmental performance. However, the effective use of such high-strength concrete and reducing the sizes of members while adopting rational structural types makes it possible to offset the negative aspects concerning environmental load per unit volume of concrete. Thus it is essential to adopt a comprehensive approach based on both the performance of concrete as a material and that required by the structure. In other words, a decrease of environmental performance per unit volume should not undermine the incentive to develop high-performance concrete. Naturally it is vitally important to greatly reduce the environmental load per unit volume of concrete at its material stage. This reveals how diverse concrete sustainability is and how great its potential. This chapter introduces technologies for sustainability related to concrete and its structures.

5.2 Concrete

5.2.1 Manufacture of Cement

5.2.1.1 Energy Consumption and Carbon Dioxide Emissions

Cement is the key ingredient of concrete, and is made through the high temperature calcination of raw materials such as limestone, clay, silica stone, etc. Approximately 1.1 tonnes of limestone, 0.2 tonnes of clay and 01~0.2 tonnes of other materials are needed to produce 1 tonne of cement. Coal and other waste materials are used for calcinations, but the ash generated by burning these materials is used as raw material for clinker, meaning that no solid wastes are discharged during the process of cement production.

However, it is unavoidable for cement production to emit large amounts of carbon dioxide (CO_2) from the limestone and the coal used for fuel. In addition, the processes of raw material proportioning and crushing and clinker crushing after adding gypsum require the consumption of electricity. Figures 5.1 through 5.3 show respectively the energy unit consumption, energy-derived CO_2 emission intensity, and process-derived CO_2 emission intensity in Japan. The recent increase in specific energy unit is due to the increasing ratio of

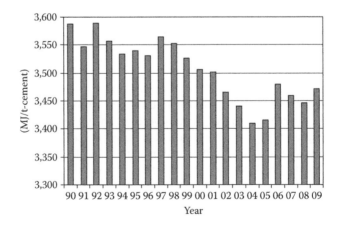

FIGURE 5.1
Energy unit consumption in Japan (*Source:* Japan Cement Association).

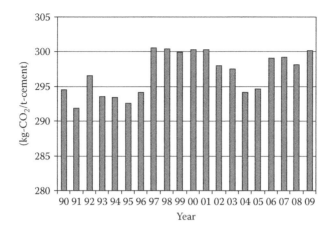

FIGURE 5.2
Energy-derived carbon dioxide emission intensity in Japan (*Source:* Japan Cement Association).

thermal in-house power generation, while the decrease in process-derived CO_2 emission intensity since 2007 is due to increasing use of blended cement.

Figures 5.4 and 5.5 show the comparisons of energy consumption and CO_2 emission intensity per tonne of clinker and cement, respectively. These data are not based on the same conditions, as different boundaries may have been used. However it is clear that there are in general significant differences between countries. This means that CO_2 emissions can be reduced substantially by raising energy efficiency. The reason Japan demonstrates outstanding energy efficiency is because it excels in energy-saving technology regarding the crushing of raw materials and use of fuel, burning, and the crushing of clinker.

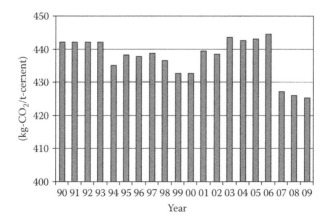

FIGURE 5.3
Process-derived carbon dioxide emission intensity in Japan (*Source:* Japan Cement Association)

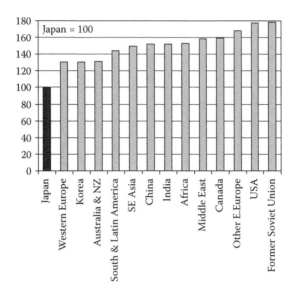

FIGURE 5.4
Energy consumption per tonne of clinker (*Source:* Japan Cement Association).

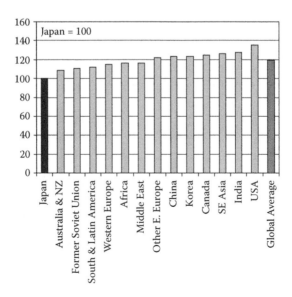

FIGURE 5.5
Carbon dioxide emissions per tonne of cement (*Source:* Japan Cement Association).

The fluidized bed cement kiln system was developed in Japan as a new cement production technology (Hashimoto and Watanabe, 1999) that reduces CO_2 by 10 to 25% compared with rotary kilns under certain conditions. NOx emissions can also be reduced by more than 40%. This technology is characterized by an autogenous hot granulation system and high energy efficiency stemming from the use of a fluidized bed. It is most effective to introduce the technology at the time of cement plant establishment or renewal.

5.2.1.2 Utilization of Wastes and By-Products

One of the characteristics of recent cement is that it uses wastes and by-products as raw materials and fuels for its production. Table 5.1 shows the main wastes and by-products used as well as their utilization purposes.

According to Japanese statistical data for 2008, approximately 38% of blast furnace slag was used as raw material for cement and 11% as supplementary cementitious material (SCM), while 64% of coal ash was also used as raw material for cement. Approximately 450 kg of wastes and by-products were used to produce 1 tonne of clinker, of which blast furnace slag and coal ash accounted for 55%. As wastes and by-products containing the key ingredients of cement such as Ca, Al, Si, and Fe, can be used as raw materials for clinker, cement production can be categorized as an environmental industry that promotes the recycling of resources.

5.2.1.3 Development of New Cement

Cement production involves the emission of large amounts of CO_2 as long as it uses limestone as a key material. It is therefore necessary to develop

TABLE 5.1

Utilization of Wastes and By-Products in Cement Production

Main Wastes and By-Products	Raw Material	Thermal Energy	SCM
Blast furnace slag	O*	–	O
Coal ash	O	–	O
Sewage sludge	O	–	–
Construction waste soil	O	–	–
Nonferrous slag	O	–	–
Wood chips	O	O	–
Waste plastics	–	O	–
Waste tires	O	O	–
Others	O	O	–

Purpose of Use spans Raw Material, Thermal Energy, and SCM columns.

* O indicates "used."

cement which uses materials other than limestone to reduce CO_2 generated from its raw materials. One of them is the cement based on belite, calcium sulfoaluminate, and calcium alumino-ferrite that has been developed by Gartner et al. (Gartner and Quillin, 2007; Walenta and Gartner, 2007; Wang et al., 2010). This cement achieves lower CO_2 emissions by reducing the amount of limestone while partially increasing gypsum, clay, and bauxite. In addition, belite-based cement allows lower calcination temperature, thereby reducing CO_2 emissions.

Sui Tongbo et al. (2007) studied the CO_2 reduction effect of belite-based cement and concluded that it allowed the calcination temperature to be reduced by 100°C, which represents 20% less energy consumption, resulting in 10% less CO_2.

Another cement has been developed which uses magnesium mixtures instead of calcium carbonates. This cement is described as carbon-negative cement as it allows the use of biomass as a fuel, while the creation of magnesium carbonates from magnesium silicates involves the absorption of CO_2 (Novacem, 2011).

As introduced above, various innovative cement production technologies have been developed as a means of reducing CO_2, and transcend the scope of current mainstream Portland cement. Because they face challenges regarding the procurement of raw materials, costs and performance, these new cements have not yet reached the point of being widely usable; however, it is vitally important to continue developing novel types of cement as alternatives to conventional Portland cement.

5.2.1.4 CCS Technology and Its Applications to Cement and Concrete

The carbon dioxide capture and storage (CCS) technology that captures CO_2 emitted at power stations and industrial plants and stores it underground is attracting attention today as a trump card concerning CO_2 management. The IPCC Fourth Periodic Assessment Report (2007) recommends introduction of the CCS system in the energy industry as well as cement and steel industries as an example of sectoral mitigation technology.

Injecting CO_2 into an oil field with low oil pressure will contribute to an increase of crude oil production. In CCS technology, it is necessary to separate CO_2 from combustion waste gas, and press it deep underground. This means that in addition to the cost of providing transportation infrastructure from the original generation point of CO_2 to the place of its storage, the technology will face environmental risks including impacts on underground water and storage stability. It is therefore anticipated that considerable time will be needed before this technology is generalized.

The establishment of CCS technology can provide one option for CO_2 emission control; however, placing too much expectation on it will become an impediment to other innovative technological developments regarding CO_2 reduction. It will be necessary in the future to give in-depth consideration to the effectiveness of CO_2 reduction through the introduction of CCS technology to cement production.

Limestone originates in sediments formed over many years by combining calcium ion dissolved in water and CO_2. In other words, in theory, if CO_2 can be captured and sufficient calcium ion is present, limestone can be obtained by their chemical reaction, which will enable us to fully circulate the CO_2 resulting from cement production. However, in order to realize this, a supply of calcium is essential. Before the existence of living organisms, CO_2 must have been fixed by inorganic chemical reactions.

Today, CO_2 and Ca are captured by living organisms such as foraminifera and coral polyps in addition to inorganic matters that form calcium carbon, thereby efficiently fixing CO_2. If such cycling can be achieved artificially, the cement industry will be able to realize carbon neutrality with respect to raw material-derived CO_2, that is to say, CO_2 generated by the decarbonation of limestone combines with Ca, and regenerates new limestone.

Although this requires Ca as the source, at least CO_2 will be circulated. A technology called mineralization via aqueous precipitation (MAP; Constantz et al., 2010) has been developed to capture CO_2 by passing the CO_2 emitted from thermal power generation through seawater, thereby combining CO_2 with the Ca contained in seawater. The aim is to use the resulting calcium carbonate as a supplementary cementitious material (SCM). The MAP process currently focuses on the CO_2 from thermal power generation, but it could of course be applied to the treatment of CO_2 generated during cement production.

Further, if SCMs produced by other industries could be used as cement substitutes, it would allow additional offsetting of captured CO_2 to that reduced in cement production. Presently studies are being carried out concerning such calcium carbonate for use as an SCM, but it can also be used as a cement material. This technology that simulates the natural cycling of Ca and CO_2 by artificial means is indeed exciting, but its success depends on the supply of Ca and the efficiency of capturing CO_2. Another concern is that extracting vast amounts of Ca from seawater may cause an environmental problem.

On the other hand, hardened cement paste has a property to chemically capture CO_2, a reaction known as carbonation. If the amount of captured CO_2 can be predicted in relation to the life of a concrete structure, such amount can be offset from the CO_2 involved in concrete production, despite the time lag between emission and capture. Moreover, if CO_2 can be forcibly absorbed in the fine powder of demolished concrete, the resulting powder with absorbed CO_2 can be used as a substitute for limestone, again despite the time lag between emission and absorption.

A new type of concrete with zero or less CO_2 emission has been developed in Japan. It is produced by using, in addition to the conventional materials (water, cement, and aggregate), a special chemical admixture that hardens concrete by reaction with CO_2, together with coal ash discharged from thermal power stations and cures the resulting concrete with CO_2 contained in the effluent gas from thermal power stations. This was realized through coal ash substitution and CO_2 capture during the process of curing. Such concrete has been used for paving blocks. This technology is introduced in Chapter 6.

Therefore, the concept of carbon dioxide capture and storage (CCS) can be rationally applied to concrete. Namely, CO_2 capture and storage can be incorporated in the processes of concrete production and demolishing. The problem is that the amount of CO_2 emitted by cement production greatly exceeds the storing capacity of concrete. It is therefore necessary to aim for the achievement of carbon neutrality by building processes that capture such emitted CO_2 in various forms while reutilizing it as a cement raw material.

5.2.2 Production of Concrete

5.2.2.1 Use of SCM

Industrial byproducts such as blast furnace slag and fly ash have been used as supplementary cementitious materials (SCMs) for the production of concrete, and this has been primarily based on a pursuit of the effective use of industrial by-products. Although these materials are usable as cement substitutes, their use is limited because their performance does not match that of Portland cement. Despite this, as it is evident that their utilization as a cement substitute contributes to the reduction of CO_2 emission, the use of industrial by-products is receiving a great deal of attention today as an interim trump card concerning CO_2 reduction.

Figure 5.6 shows world crude steel production in 2010. The total volume is approximately 1.41 billion tonnes, of which China accounts for approximately 44%. Approximately 485 million tonnes of steel slag are assumed to be discharged from steel production (1.41 billion tonnes × 0.344); in other words, nearly 500 million tonnes of slag can be used as resources. The discharge factor 0.344 applied here was obtained by dividing the total volume

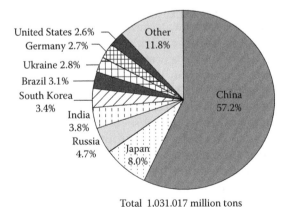

Total 1,031.017 million tons

FIGURE 5.6
World crude steel production in 2020. (*Source:* World Steel Association.)

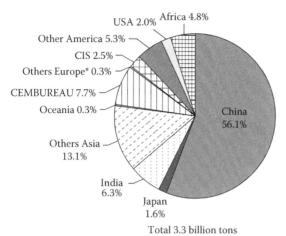

FIGURE 5.7
World cement production in 2010 by country and region. (*Source:* CEMBUREAU.)

of blast furnace slag and steel slag (or converter slag and electric arc furnace slag) by the volume of crude steel in Japan. Steel slag is used as a material for blast furnace slag cement that enables direct CO_2 reduction, aggregate for concrete, road beds, and aggregate for asphalt mixtures.

Figure 5.7 shows world cement production in 2010 by country and region. The total volume is approximately 3.3 billion tonnes, of which China accounts for approximately 56.1%. It is not easy to calculate the amount of CO_2 emitted from cement production. This is because, in addition to the fact that specific CO_2 emissions from cement production are likely to vary significantly, the ratio of SCMs (limestone powder, pozzolans, blast furnace slag, fly ash, etc.) other than clinker and the amounts of CO_2 contained in SCMs are unknown.

P.K. Mehta (2009) estimated that Portland cement contains 84% clinker and that the CO_2 emission intensity of clinker production is 0.9 (tonnes CO_2/tonnes cement). The average CO_2 emission intensity of countries and regions investigated by Humphreys et al. (2002) is 0.87 (tonnes CO_2/tones cement). The 2008 clinker/cement ratio in Japan is estimated as 0.9. Based on the above data, calculations were made by assuming that CO_2 emission intensity is 0.9 (tonnes CO_2/tonnes cement), and clinker/cement ratio 0.9, as CO_2 emission was estimated to be on the larger side here. CO_2 emissions involved in SCMs other than clinker were disregarded for the calculation. The result obtained is approximately 2.67 billion tonnes ($3.3 \times 0.9 \times 0.9$) of CO_2 emissions from total world cement production.

If 2.97 billion tonnes (3.3×0.9) of cement clinker is replaced by the entire 480 million tonnes of steel slag, approximately 16% of clinker will be reduced, representing a 432 million tonne reduction of CO_2. However, as blast furnace slag is used for cement substitute slag, the actual figure will be lower.

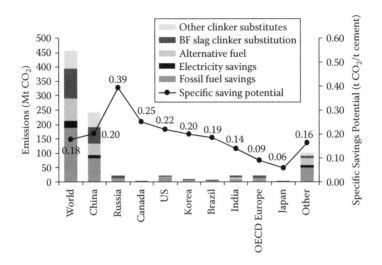

FIGURE 5.8
Potential carbon dioxide reduction by global cement industry (*Source:* IEA).

Meanwhile, world electricity generated by coal-fired thermal power in 2008 was 8,262,523 GWh (IEA, 2008a), accounting for 40% of total electricity. According to the authors' investigation, in general, 0.0354 kg/kWh of fly ash is generated, of which 40% is thought to be usable as a cement substitute. Based on this, fly ash available globally will be approximately 117 million tonnes.

If this amount were used to replace the 2010 world cement production of 3.0 billion tonnes, the use of cement could be reduced by approximately 3.9%. If the amount of fly ash available were doubled, it would amount to 234 million tonnes, reducing the use of cement by 7.8%. In this case, with blast furnace slag and fly ash combined, 23.8% of cement substitution would be possible. However, as these estimations are considerably high, it is safer to assume that the reduction in reality will be in the range of 15 to 20%.

Since replacement at such level will have little impact on concrete performance, compared with performance using ordinary Portland cement only, it is important to make maximum use of these SCMs on a global scale in the future. As shown in Figure 5.8, the IEA (2008b) has, by considering diverse reduction factors, estimated the CO_2 reduction potential of the global cement industry as 450 million tonnes.

With respect to the environmental burden of blast furnace slag cement and fly ash—both by-products— it will be better to disregard the burden resulting from their production and rather encourage raising the incentive for their reuse. Otherwise, disposal of these materials will be a huge problem and their releasers will benefit greatly from such reuse. However, it is perhaps better for users to consider the burden of transporting them from the place of their release to where they are reused.

5.2.2.2 Use of Chemical Admixtures

Among the various types of chemical admixtures, the most commonly used are AE water-reducing admixtures and AE high-range water-reducing admixtures. There are plant-based and chemical-based raw materials used for chemical admixtures. For example, lignosulfonates, by-products of wood pulp production, are plant-based. Polycarboxylate ethers used for high-range AE water-reducing admixtures are chemical-based synthesized materials.

Therefore the embodied energy of chemical admixtures differs greatly, depending on their raw material; however, as their dosages are generally small compared with other materials making up concrete, their embodied energy can be ignored. Table 5.2 shows the CO_2 emission intensity of chemical admixtures surveyed by the Japan Concrete Institute. The yearly consumption of chemical admixtures is known to be 1% that of cement and is approximately 30 million tonnes. The actions of chemical admixtures are as follows:

Air entrainment

Reduction of mixing water content

Increase of workability

Improvement of durability

Others

A decrease in the unit water content through the use of water-reducing admixtures contributes to the conservation of water resources. The amount of water reduced varies, depending on the type of admixture and quality of aggregate used. If it is assumed that 8 billion cubic meters (20 billion tonnes) of concrete are produced and the unit water content is reduced by 10%, making it 150 kg through the use of admixtures, 17 kg of water per cubic meter will be conserved, saving 136 million tonnes in total. For comparison, the global consumption of mineral water is reportedly 126 million tonnes.

Reducing the unit water content allows reduction of the unit cement content, which ultimately contributes to lower CO_2 emissions from cement production. Table 5.3 (Sakai, 2009) shows the estimated CO_2 reduction effect

TABLE 5.2

Unit-Based CO_2 Emissions of Chemical Admixtures

Constituent	CO_2 (kg CO_2/t)
Lignosulfonate	100–150
β-naphthalenesulfonate	150–250
Melaminesulfonate	50–150
Polycarboxylate ether	100–350

TABLE 5.3

Carbon Dioxide Reduction Effect through Use of AE High-Range Water-Reducing Admixture

Concrete Type	W/C (%)	s/a (%)	Unit Content (kg/m³)/CO₂ Emission (kg CO₂/m³)					CO₂ Reduction when Changed to AE High-Range Water-Reducing Admixture
			W	C	S	G	Ad	
AE water-reducing admixture	50	46.6	180	360 (276.0)	801 (3.0)	935 (2.7)	0.93 (0.1)	
AE high-range water-reducing admixture	50	47.8	170	340 (260.6)	842 (3.1)	935 (2.7)	3.5 (0.8)	14.6 kg CO₂/m³

through the use of chemical admixtures. The total amount of CO_2 emitted during the production of ingredient materials for concrete with AE water reducing admixtures and AE high-range water-reducing admixtures was calculated, and the environmental load-reducing effect by AE high-range water-reducing admixtures was estimated. This reveals that changing the use of AE water-reducing admixtures to AE high-range water-reducing admixtures will allow reduction of cement by 5.6%, resulting in 14.6 kg less CO_2 emission per cubic meter in the entire mixture proportion. Thus, AE high-range water-reducing admixtures can be regarded as one of the tools for CO_2 reduction in concrete.

The use of chemical admixtures can enhance the durability of concrete. Air-entraining admixtures increase resistance to freezing and thawing, while AE high-range water-reducing admixtures produce high-strength concrete, which provides better durability. Other admixtures such as shrinkage reducers and corrosion inhibitors are also effective for increasing the durability of concrete. This contributes to the longer life of concrete structures, while at the same time enhancing the effective use of resources and environmental load reduction. Hydration-controlling admixtures, by properly controlling returned concrete and concrete wash water, can reduce landfill waste.

5.2.2.3 Use of Recycled Aggregate

Today, because of its abundant raw materials and excellent performance, concrete is the most used material on Earth next to water. Although no statistical data concerning its output is available, when simply calculated by analogy based on a 3.3-billion tonne cement output in 2010 and assuming the unit cement content as 0.3 t/m³, it amounts to 27.5 billion tonnes. The entire amount of cement is not necessarily used for the production of concrete, but at least 20 billion tonnes of concrete are likely to be produced annually worldwide.

On the other hand, concrete structures are demolished for various reasons. According to a survey by the Cement Sustainability Initiative (CSI) of the World Business Council for Sustainable Development (WBCSD 2009), 900 million tonnes of construction and demolition waste are generated annually in Europe, the United States, and Japan. Assuming that two-thirds of this is concrete, it accounts for 6% of concrete output.

The benefits of recycling demolished concrete include reduction of landfill waste and fewer landfill sites, virgin resource substitution, environmental cost reduction due to natural resource mining, and disposal cost reduction as landfill tax.

The largest proportion of demolished concrete is used for road bedding. In Japan 96% or more is used as recycled aggregate for road bedding; however, there are signs of a decline in such application due to reduced road construction recently. Although the use of recycled aggregate is extremely important in terms of resource-saving, its application to concrete is naturally limited as its quality varies. Australia uses up to 30% of recycled aggregate, and concrete containing recycled aggregate is commercialized and used in Germany, Switzerland, and Austria (WBCSD Cement Sustainability Initiative, 2009). In Japan, the following three standards have been established concerning recycled aggregate.

JIS A 5021 – Recycled aggregate for concrete – Class H

JIS A 5022 – Recycled concrete using recycled aggregate – Class M

JIS A 5023 – Recycled concrete using recycled aggregate – Class L

Table 5.4 shows definitions regarding the oven-dry density and water absorption ratio of recycled aggregate under these classifications. The definitions of Class H are the same as those of ordinary aggregate. The use of recycled aggregate usually requires confirmation of alkali–silica reactivity, but this can be avoided by using blended cement. Table 5.5 shows the upper limits of the impurity content in Class H recycled aggregate.

TABLE 5.4

Oven-Dry Density and Water Absorption of Recycled Aggregate (Japanese Industrial Standard)

	Class H		Class M		Class L	
	Coarse	Fine	Coarse	Fine	Coarse	Fine
Oven-dry density (g/cm^3)	Not less than 2.5	Not less than 2.5	Not less than 2.3	Not less than 2.2	–	–
Water absorption (%)	Not more than 3.0	Not more than 3.5	Not more than 5.0	Not more than 7.0	Not more than 7.0	Not more than 13.0

TABLE 5.5

Upper Limits of Impurity Content in Class H (Japanese
Industrial Standard)

Category	Impure Substance	Limit (mass%)
A	Tile, brick, ceramic, asphalt concrete	2.0
B	Glass	0.5
C	Plaster	0.1
D	Inorganic substances other than plaster	0.5
E	Plastics	0.5
F	Wood, paper, asphalt	0.1
Total		3.0

Technologies for recycled aggregate production in Japan include the heating and rubbing (Tatebayashi et al., 2000), eccentric-shaft rotor (Yanagibashi et al., 1999), and mechanical grinding (Yoda et al., 2004) methods. In the heating and rubbing method, concrete masses are heated at 300°C and the cement paste content is weakened to remove mortar and cement paste from the aggregate. Figure 5.9 shows an overview of a recycled aggregate production system using this method. Figure 5.10 illustrates the recycled coarse and fine aggregate produced by the system. While the production of recycled aggregate generated a large amount of fine powder, it also indicated the possibility of using such fine powder as a substitute solidification material for the deep mixing stabilization method (soil cement walls) (Uchiyama et al., 2003).

In the eccentric-shaft rotor method, crushed concrete lumps are passed downward between outer and inner cylinders that eccentrically rotate at a high speed to separate the concrete into coarse aggregate and mortar through the grinding effect. Figure 5.11 presents an overview of a recycled

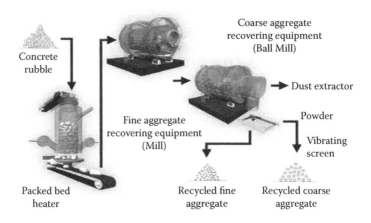

FIGURE 5.9
Recycled aggregate production system (Mitsubshi materials).

FIGURE 5.10
Recycled coarse and fine aggregate production by heating and rubbing recycling method.

aggregate production system using this method. Mechanical grinding is a method used to produce coarse and fine aggregates by separating a drum into small sections with partitions and loading the drum with iron balls for grinding and rotating the partitions. Figure 5.12 shows an overview of the recycled aggregate production system using this method. The coarse aggregate produced by these methods has been used for construction projects.

The production of recycled aggregate requires the use of energy. Embodied energy therefore differs, depending on the methods by which recycled aggregate is produced. Table 5.6 shows the CO_2 emissions in the production of various aggregates (Architectural Institute of Japan, 2008). As the heating and rubbing recycling system demands a large amount of energy for

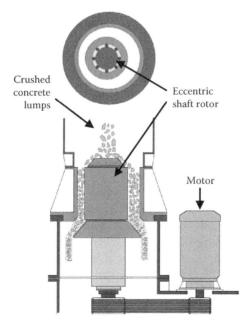

FIGURE 5.11
Recycled aggregate production via eccentric-shaft rotor method (Takenaka Corporation).

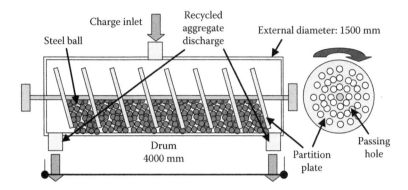

FIGURE 5.12
Recycled aggregate production via mechanical grinding method (Kajima Corporation).

TABLE 5.6

CO_2 Emissions in Production of Various Aggregates

Aggregate Category	Unit-Based CO_2 (kg CO_2/t)
Crushed stone	2.9
Crushed sand	3.7
Recycled aggregate (eccentric-shaft rotor recycling)	3.7
Recycled aggregate (heating and rubbing recycling)	41.5
Molten slag	2293.6
Artificial aggregate (fly ash)	50.0
Artificial aggregate (expansive shale)	47.0

heating, the CO_2 emission intensity is very high, while the CO_2 emission intensity by the eccentric-shaft rotor recycling system is low, provided that the latter produces aggregate of inferior quality compared with that of the former. Thus, it is necessary to be aware that the relation between an aggregate's quality and embodied energy differs greatly, depending on the state of the original concrete and production method.

5.2.2.4 Use of By-Product Aggregate

By-product aggregate includes steel slag aggregate, copper slag aggregate, and molten slag aggregate. Steel slag can be divided broadly into blast furnace slag and steelmaking slag. Blast furnace slag is generated when iron ore is melted and reduced. In other words, it consists of a compound of limestone combined with the non-ferrous ingredients in iron ore such as silica, and ash contained in coke, and is produced at a rate of around 290 kg per tonne of pig iron (Nippon Slag Association, 2011). Slag discharged from a blast furnace, which is in a high-temperature molten state, is subjected to natural cooling or sprinkling of water or to rapid high-pressure cooling. The former treatment produces crystalline, rock-like air-cooled slag, while the latter produces glassy granulated slag.

Steelmaking slag, meanwhile, is a by-product of steel production using pig iron and scrap iron. In steelmaking slag, there are converter slag and electric arc furnace slag. Converter steel slag, which is produced by the slow cooling of slag discharged from a converter, is produced at the rate of approximately 110 kg per tonne of converter steel (Nippon Slag Association, 2011). Electric arc furnace slag, which is produced when scrap iron is melted and refined, is produced at the rate of approximately 70 kg of oxidizing slag and 40 kg of reducing slag respectively per tonne of electric arc furnace steel (Nippon Slag Association, 2011).

It is well known that blast furnace slag is used as a raw material for cement, but blast furnace slag and electric arc furnace oxidizing slag are also used as aggregate, and Japan has established relevant standards for them. Apart from these uses, steel slag has a wide range of applications, for instance as cement clinker raw material, road bedding, and backfill material for revetments.

When copper is refined from ore, copper slag is generated. A major feature of copper slag is that it has a specific gravity of around 3.5, which is an obstacle to its use as concrete aggregate. This feature is, however, turned to advantage in its use as aggregate for heavy-weight concrete. In Japan, despite the established standards on copper slag for use in concrete aggregate, copper slag is not widely used for this application. The main applications of copper slag include as cement raw material, caisson filling, interlocking block aggregate, and blasting sand.

One problem of contemporary society is the increase in waste. Until now, waste material has been treated at a final disposal site, but it has become difficult to secure locations for such facilities. Incineration has consequently become normal as a way of reducing the volume of waste, but as incineration ash contains dioxins and heavy metals, detoxification treatment is required. This involves melting the incineration ash at a temperature of 1300°C or above, which produces a residue of molten slag. In Japan, there are agreed standards on concrete aggregate made with molten slag, but as the use of this aggregate is not yet accepted under ready-mixed concrete (RMC) standards, a large part of the molten slag is used in concrete products. An exception in Japan is Kagawa Prefecture, where molten slag discharged in the melting of illegally dumped industrial waste is used as raw-concrete aggregate for unreinforced concrete structures.

In this way, the use in the field of concrete and construction of the various types of slag occurring as by-products of metal refining can make a major contribution to resource and energy conservation and the realization of low carbon environments, and an expansion in their use would be welcome. However, molten slag is generated by high energy consumption for the treatment of waste alone. Therefore, the use of molten slag in concrete is not necessarily a welcome prospect, but should rather be seen as an unavoidable necessity. The essential point is that it is important to reduce waste to the minimum.

5.2.2.5 Recycling of Sludge Water and Recovery Aggregate

At a ready-mixed concrete plant, the washing of concrete mixers and agitating trucks generates sludge water. Sludge water contains sludge solids in the form of hydrates and aggregate powder. The recycling of sludge water as concrete mixing water is important from an environmental viewpoint, but sludge water with a high solid content is known to have a negative effect on the properties of the concrete. In Japan, however, the standard allows sludge water to be used as long as the sludge solid content is guaranteed not to exceed 3%.

It is possible to recycle the aggregate recovered by water treatment of the mortar and small amounts of residual concrete that adhere to the interiors of concrete mixers and agitating trucks. However, according to surveys in Japan, less than half of such material appears to be recycled as concrete aggregate, with more than half used as ballasting. As Japanese standards contain no regulations on the re-use of recovered aggregate, where it is recycled in concrete, the resulting concrete will be regarded as off-specification material.

5.2.3 Down Cycling of Residual and Returned Concrete and Sludge Cake

Residual concrete and returned concrete are generated at ready-mixed concrete plants. The treatments applied to these materials include treatments to produce masonry units and other concrete products; treatments to allow the concrete to be reused involving long-term storage after use of a special chemical admixture to interrupt cement hydration, which process can be reactivated before re-use; and treatments whereby the concrete is hardened, crushed, and recycled as road bedding.

When manufacturing concrete products, it is necessary to provide equipment and systems that minimize energy consumption. Additionally, the balance of demand and supply and transportation distance are important factors in determining the effective use of the products.

The sludge water resulting from washing the agitating trucks that are used to mix and transport concrete is separated using a dehydrating device into filtered water and dehydrated cake, and the filtered water is reused as sludge water for mixing. The dehydrated cake undergoes treatment as industrial waste material or is recycled as road bedding, etc., after being mixed with cement, hardened, and crushed.

The generation of residual concrete, returned concrete, and sludge cake is unavoidable, but as their treatment requires additional energy, systems need to be developed to minimize the amounts generated.

5.2.4 Renewable Energy for Concrete Production

Concrete production generally makes use of commercial electric power. Table 5.7 shows worldwide electricity production by method of generation (IEA, 2011). The renewable energy with the largest share of electricity

TABLE 5.7

Worldwide Electricity Production by Method of Generation, 2008

World Electricity Production (%)							
Coal	Oil	Gas	Nuclear	Hydro	Biomass	Wind	Other
40.8	5.5	21.2	13.5	16.2	1.0	1.0	0.8

generation is hydroelectric power, with over 16%, while other renewable energies account for 3%. In world terms, just under 20% of electricity is thus generated by renewable energy; assuming that the average figure applies to concrete production, this means that nearly 20% of the electric power it uses is based on renewable energy. However, as the energy sources for worldwide electricity production vary widely between countries and regions, simple arguments do not apply.

To generate the electric power needed for concrete production using renewable energy, it would be necessary for the industry to install relevant electricity-generating plants, but so far there have been few examples of such moves.

5.2.5 Transport of Constituent Materials and Concrete

Concrete production requires the transportation of cement, aggregate, supplementary cementitious material (SCM), and other raw materials. Transportation is also required to bring the produced concrete to the construction site. Both processes require fuel for vehicle transportation. Fuel consumption generates CO_2, NOx, SOx, and other emissions. Concrete raw materials should therefore ideally be procured from as near a source as possible. Also, ready-mixed concrete should be supplied from a plant near the construction site.

Figure 5.13 shows the results of a survey of CO_2 emissions per cubic meter of concrete except for concrete materials in a certain region of Japan. The data

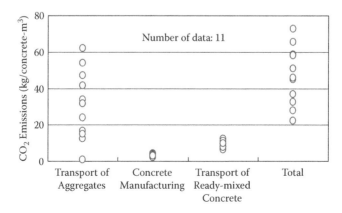

FIGURE 5.13

Results of survey of carbon dioxide emissions per cubic meter except for concrete materials,

show a wide variation in the levels of CO_2 emissions arising from materials transport. If one cubic meter of concrete requires 300 kg of ordinary Portland cement and the CO_2 emission factor is 0.8, 240 kg of CO_2 are emitted, with CO_2 emissions arising from material transport accounting for up to 25% of the emissions related to cement. In contrast, there is no major variation in the level of CO_2 emissions arising from concrete production or the transport of ready-mixed concrete and the absolute emission volumes are also small. As transport distance thus has a major influence on the carbon footprint of concrete, the sourcing of materials needs to be taken into account.

5.3 Concrete Structures

5.3.1 Utilization of Thermal Mass

With recent global warming, the air-conditioner utilization ratio has been constantly rising, resulting in an increase of CO_2 emissions every year. Instead of pursuing the comfort of interior space through the use of mechanical air-conditioning and lighting systems that consume energy, there is another system called passive design that achieves the effects of heating, cooling, and lighting, through the use of the flow of natural heat and light.

With respect to the utilization of sunlight, an active solar system uses sunlight as the heat source of machinery and equipment including photovoltaic generation and solar water heating facilities, and a passive solar system assigns the heat transfer to natural convection, conduction, and radiation, without using machinery and equipment for the heating processes of collection, conservation, and radiation of sunlight. In a passive solar system, as shown in Figure 5.14, a heat saving insulator is installed inside the building in winter, in order to gain heating for use during the evening and nighttime or on cloudy days, by utilizing sunlight during the daytime.

When considering the amount of heat storage alone, the index to assess the adequacy of a heat storage insulator can be the amount of heat required to raise the temperature of the material by ΔT ($\rho C \Delta T$ where ρ = density and C = specific heat). It is nevertheless desirable for the insulator to consist of a material that does not easily cool once it is heated. The amount of heat storage in the insulator over t hours is expressed as follows.

$$Q \propto \sqrt{t} \; T \frac{\lambda}{a^{1/2}}$$

where λ = thermal conductivity and a = thermal diffusivity (= $\lambda/\rho C$). This illustrates that a material with high a value is not suitable for an insulator, while one with a high $\lambda/a^{1/2}$ value and with a moderate a value is suitable.

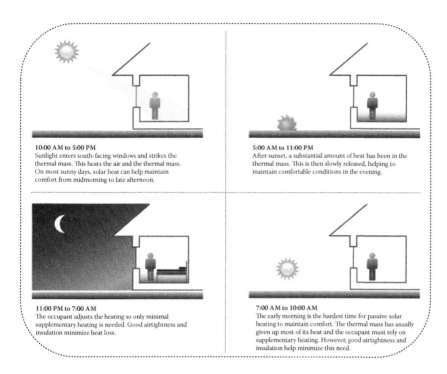

10:00 AM to 5:00 PM
Sunlight enters south-facing windows and strikes the thermal mass. This heats the air and the thermal mass. On most sunny days, solar heat can help maintain comfort from midmorning to late afternoon.

5:00 AM to 11:00 PM
After sunset, a substantial amount of heat has been in the thermal mass. This is then slowly released, helping to maintain comfortable conditions in the evening.

11:00 PM to 7:00 AM
The occupant adjusts the heating so only minimal supplementary heating is needed. Good airtightness and insulation minimize heat loss.

7:00 AM to 10:00 AM
The early morning is the hardest time for passive solar heating to maintain comfort. The thermal mass has usually given up most of its heat and the occupant must rely on supplementary heating. However, good airtightness and insulation help minimize this need.

FIGURE 5.14
Passive solar storage system.

Having a large thermal capacity and relatively high thermal conductivity, concrete is a suitable material for a heat storage insulator. In order to obtain full benefit from such heat storage function, it is recommended to finish its surface as exposed, tiled, painted, or stucco plastered, while finishes using carpets and gypsum plaster board reduce the effect. As shown in Figure 5.15, in the case of detached houses, CO_2 emissions from concrete buildings are higher until completion of their construction compared with wooden buildings, but as the use of an air-conditioner can be reduced through good use of the heat storage effect, it has been estimated that concrete buildings eventually generate fewer emissions after several to a little over ten years.

There are some types of concrete that are used as a heat storage insulators in passive solar system as described below. In all cases, the concrete used for walls and floors serves as an insulator.

5.3.1.1 Direct Gain System

As shown in Figure 5.16a, this system, having concrete indoor floors and walls, is designed to introduce sunlight from south facing windows and to store the heat in concrete during the daytime. The concrete begins to radiate

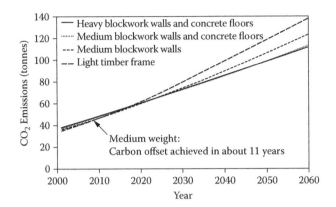

FIGURE 5.15
Comparison of carbon dioxide emissions from concrete and wooden buildings.

the stored heat from its surface as the temperature drops during the night, thereby achieving the heating effect for the house.

To increase heat storage efficiency, it is desirable to maximize the area of concrete that absorbs sunlight, while using a finishing material with a high sunlight absorption rate. The effective thickness of concrete for heat storage and radiation for a one-day cycle is 15 to 20 cm, and it is more effective in terms of gaining heat storage to increase the surface area rather than the thickness of concrete.

5.3.1.2 Trombe Wall System

As shown in Figure 5.16b, this system, which involves the installation of a concrete wall inside sunlight collector windows, achieves the heating effect by the heat transmitted through the wall. The thicker the wall, the more heat is stored and the longer the time lag of heat transfer. On the other hand, the attenuation of transferred heat becomes greater.

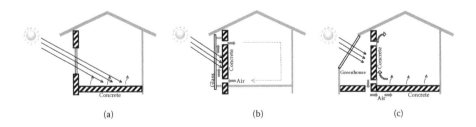

FIGURE 5.16
(a) Direct gain system. (b) Trombe wall system. (c) Attached solar greenhouse system.

The thickness of concrete for this system is generally 20 to 30 cm and the heat transfer time lag is about six to eight hours. The heat storage and radiation for a one-day cycle is 15 to 20 cm, and it is more effective in terms of heat storage to increase the surface area rather than the thickness of concrete. In some cases, the system provides air vents at the top and bottom of the concrete wall to accelerate heat transfer by way of air convection. Heat collection efficiency has been considerably improved by the development of a construction method to fill the layer of air between the window glazing and concrete wall with transparent insulator material.

5.3.1.3 Attached Solar Greenhouse System

As shown in Figure 5.16c, this system, by designating a greenhouse attached to the south side of the living area as a heat collection unit, sends the air heated in the greenhouse to the underfloor of the house where concrete slabs are installed.

Meanwhile, in the warm climate (summertime), concrete can be utilized as an important element for the passive cooling system. In this case, the heat absorbed by and stored in the concrete floor slabs during the daytime is cooled by the outdoor air introduced by ventilation at night, resulting in a stable indoor temperature throughout the day, as shown in Figure 5.17. Through this cooling system, energy consumption arising from the daytime use of air-conditioning becomes less, thereby reducing CO_2 emissions. However, it is still necessary to prevent sunlight from entering the rooms in summer and for its heat to be stored in the concrete floor slabs, while the installation of eaves, balconies, and louvers is essential to use concrete for the passive cooling system effectively.

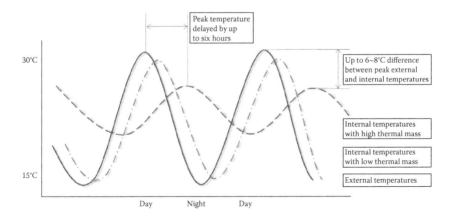

FIGURE 5.17
Heat storage effect by passive cooling systems.

FIGURE 5.18
Exposed flat slab ceiling plus natural ventilation.

With respect to the use of concrete floor slabs as heat storage insulators in the passive cooling system, there are various methods based on the floor slab configuration and ventilation system, as follows.

Exposed flat slab ceiling + natural ventilation — As shown in Figure 5.18, this is a method to cool flat or uneven floor slabs through the use of natural ventilation. By finishing the ceiling surface in a grid, grooved, or corrugated form, it has little effect on through radiation but doubles thermal conductivity through convection because of the increased surface area, resulting in enhancement of the heat radiation of concrete. Cooling efficiency by this method is 15 to 20 W/m² and 20 to 25W/m² in the flat slab and uneven slab, respectively.

Exposed waffle slab ceiling + underfloor forced ventilation — As shown in Figure 5.19, this is a method to cool concrete slabs by introducing fresh air into the space under the raised floor. This method is effective also in that its cooling is carried out by direct contact of the air on both sides of the concrete slab, and therefore the heat storage effect can be enhanced by making the slab thicker. In addition, by making the underfloor air flow turbulent, the cooling efficiency by convective heat transfer increases further. The cooling efficiency by this method is 20 to 30 W/m² and 25 to 35W/m² in the flat slab and uneven slab, respectively.

Exposed hollow core slab ceiling + underfloor forced ventilation — As shown in Figure 5.20, this is a method to cool concrete slabs by introducing fresh air into the continuous hollow installed in the precast concrete slab and circulating it at low speed. The temperature difference between the air passing through the hollow and the concrete slab is only 1 to 2°C. Cooling efficiency by this method is 40 W/m² in the case of the standard type.

FIGURE 5.19
Exposed waffle slab ceiling plus underfloor forced ventilation.

5.3.2 Pervious Concrete Pavement

As most pavement materials commonly used today are impervious, leaked car oil and other harmful substances remain on paved road surfaces and flow directly into rivers and lakes with rainwater, causing water contamination as well as damage to fish and other aquatic creatures. Pervious concrete, having hardly any fine aggregate, consists of a structural body composed of thick coatings of cement paste on the surfaces of coarse aggregate particles linked with each other. Formed with continuous voids with a volume ratio of 15 to 25%, it can spontaneously and easily penetrate rainwater into the ground at a speed of 200 L/m²/min, as shown in Figure 5.21.

Thus hardly any harmful substances such as fat and oil directly flow into rivers and lakes, and most of such substances absorbed by the pervious concrete are detoxified through filtration and microbial decomposition, and returned to the natural environment.

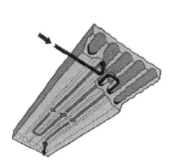

FIGURE 5.20
Exposed hollow core slab ceiling plus underfloor forced ventilation.

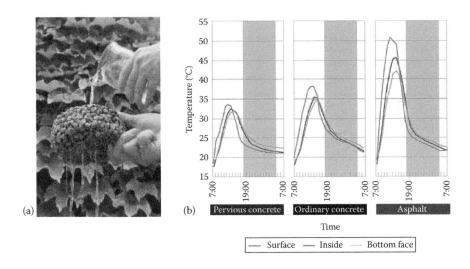

FIGURE 5.21

(a) Water penetration into pervious concrete. (b) Pavement temperature of pervious concrete and ether materials.

As the pervious concrete pavement, unlike asphalt pavement, is light in color, it absorbs less solar radiation heat, while its high void structure achieves a low heat storage effect and high water retention. As shown in Figure 5.22, it has a small temperature rise during the daytime and rapid temperature drop after sunset, also leading to possible reduction of the heat island phenomenon. Furthermore as pervious concrete allows the permeation of rainwater and air, it does not hinder the growth of trees in paved car parks and streets,

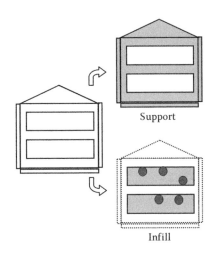

FIGURE 5.22

Support and infill in a building.

thereby contributing to mitigation of the heat island phenomenon. It also reduces the noise of car tires with its high sound proof property.

In addition, the pervious concrete pavement is not likely to cause the splashing of water as rainwater penetrates into the ground without staying on the surface. This not only provides a road surface with excellent conditions for driving and walking, but also helps decrease the risk of flooding by the rising level of underground water and water-bearing stratum, which lowers the water drainage level.

5.3.3 High Performance Concrete Structures

As cement consumption increases when the water-to-cement ratio of concrete is made smaller, CO_2 emissions from the production of 1 m^3 of concrete will increase. However this enhances the strength of concrete, making it possible for the cross section areas of columns, beams, and girders to be smaller and their intervals larger. As a result, concrete consumption for an entire building becomes smaller, resulting in lower cement consumption. Concrete with a smaller water-to-cement ratio also has higher durability, which increases the service life of concrete structures and extends the building renewal cycle. In the long term therefore, the consumption of both concrete and cement as well as waste generation can be reduced.

For example, suppose that the relation between the concrete's cement-to-water ratio (C/W) and compressive strength (F, MPa) is expressed by the following formula:

$$F = 25 \bullet (C/W) -10$$

In this case, the water-to-cement ratios of concrete with a compressive strength of 40 MPa, 100 MPa, and 200 MPa will be 0.50, 0.23, and 0.12, respectively. If the unit water content of all these concretes is 175 kg/m^3, the unit cement contents will be 350 kg/m^3, 770 kg/m^3, and 1470 kg/m^3, respectively.

Supposing that column members with the same load-bearing ability using concrete with compressive strength of 40 MPa and 100 MPa are needed, the system will require five and two times the amount of concrete with a compressive strength of 200 MPa, respectively. Thus, when based on the amount of cement required for concrete with a compressive strength of 40 MPa as one, those required for concrete with a compressive strength of 100 MPa and 200 MPa will be 0.88 and 0.84 respectively, showing that cement consumption is less when high performance concrete is used.

In terms of the density, it is 2.3 t/m^3 for concrete with a compressive strength of 40 MPa and 2.4 t/m^3 for concretes with compressive strengths of 100 MPa and 200 MPa, showing only a marginal difference. This means that compared with a case in which concrete with a compressive strength of 200 MPa is used, approximately five and two times more concrete will have to be transported when using concretes with compressive strengths of 40 MPa and 100 MPa,

TABLE 5.8

Years Required for 30-Millimeter Deep Carbonation Development and CO_2 Emissions per Cubic Meter of Concrete

Water-to-cement ratio	60%	50%	44%
Period required for carbonation to reach depth of 30mm, A (years)	47.3	162.3	679.2
Weight of cement (kg/m³)	292	350	398
CO_2 B emissions (kg/m³)	225.3	270.1	307.1
CO_2 emissions (kg/m³/year), C = B/A	4.76	1.66	0.45

Note: Unit water content of all these concretes is 175 kg/m³.

respectively. The conclusion based on the above is that using high performance concrete enables the reduction of cement consumption and concrete transportation volume, while reducing CO_2 emissions in the production of concrete.

Meanwhile, the water-to-cement ratio of concrete also correlates strongly with the mass transfer resistance of concrete. Making the ratio smaller makes it possible to reduce invasion and transfer into the concrete of substances such as carbon dioxide and chloride ion that accelerate the deterioration of reinforcing bars and concrete, thereby extending the lives of concrete structures.

Table 5.8 shows a comparison of concretes with different water-to-cement ratios. The number of years that carbonation takes to reach a depth of 30 mm (the location of reinforcing bars) and the carbon footprint that occurs up until the stage when 1 m³ of concrete has been produced were calculated to compare CO_2 emissions per year during the period that carbonation took to reach a depth of 30 mm. When the water-to-cement ratio is smaller, the mass transfer resistance of concrete increases, allowing reduction of the annual carbon footprint during the period that carbonation takes to reach a depth of 30 mm, which fully compensates for the increased amount of CO_2 emissions from the production of concrete.

To enhance concrete durability, in addition to the method of reducing the water-to-cement ratio, there is another method which applies concrete densification to prevent substances that cause deterioration from entering into concrete, by adding supplementary cementitious materials such as blast furnace slag fine powders, fly ash, and silica fume. This method serves dual purposes as it can reduce cement consumption per cubic meter of concrete.

5.3.4 Adaptable Building and Open Building

The service life of a building comprises various concepts such as physical life, functional life, social life, and economic life. Normally, as lives other than the physical life are shorter, buildings are demolished and their lives thus determined.

However there are some buildings that adapt to all changes throughout their lives without depleting their functional, social, or economic lives before their physical one. They are known as "adaptable buildings" and defined as "easily updated or modified to meet changing needs or requirements" and should have excellent "performance through flexibility." It is not possible to quantitatively assess such performance today; nevertheless, in order to produce an adaptable building, it is important to consider the following to minimize restrictions concerning future alteration in floor and sectional plans or renewal of service wiring and piping:

1. Large intervals between structural members
2. Few interior load-bearing walls
3. Simple form structural members
4. High floor height
5. Separations of structural body and service wiring and piping

The concept of the adaptable building has been presented concretely by way of an "open building" (John Habraken, 1961)—"an innovative approach to design and construction that enhances the efficiency of the building process, while increasing the variety, flexibility and quality of the product." The idea underlying this is to disassemble and separate a building in concert with the different purposes and intentions of its owner, occupant, and administrator, in other words, by clearly separating independent building components into several levels, so that any changes and renewals that occur on one level will be limited to this level, without affecting or being interfered with by other levels, thereby securing decision making-based diversity and independence.

More specifically, as shown in Figure 5.23, by dividing a building into two levels, one is the "support" level consisting of the skeleton and supporting members that make up the functional base (members with structural support function), and the other is the "infill" level that provides the occupant

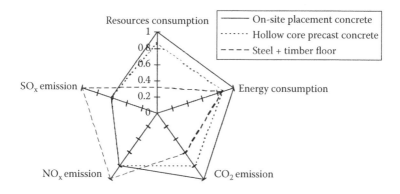

FIGURE 5.23
Comparison of environmental loads.

with functions (non-structural members, interior finishing and building services). The design assumes the life of the support level to be approximately one hundred years and the infill level to be from ten to twenty years and adaptation to various changes in future building services becomes possible, while the longer life of the building can be pursued as members, interior finishes, and services that become functionally obsolete can be easily renewed without damaging the structural members.

5.3.5 Precast Concrete Structures

When constructing concrete structures, it is often the case that the consumption of resources and energy as well as the generation of waste can be reduced by using precast concrete members instead of carrying out on-site concrete placement. Figure 5.24 shows a comparison of environmental loads between a floor slab using concrete made by ordinary on-site placement and a hollow core precast concrete shown in Figure 5.25. This demonstrates that the use of hollow core precast concrete contributes to a reduction in the amount of concrete and reinforcing bars by 40% and 50%, respectively.

In terms of energy consumption during their production process, although the hollow core precast concrete is larger than on-site placement concrete, when viewing the entire construction of the building as a whole, the use

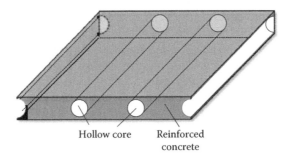

Hollow core   Reinforced
                   concrete

FIGURE 5.24
Hollow core precast concrete slab.

FIGURE 5.25
Mechanical anchorage bars.

TABLE 5.9

Comparison of Environmental Loads by Concrete Type

Parameter	Hollow Core Precast	Half Precast	On-Site Placement
Weight of concrete (kg)	263.72	423.00	423.00
Weight of reinforcing bars (kg)	3.22	6.44	6.11
Eutrophication (kg PO_4^{3-})	0.0356	0.0448	0.0410
Fossil fuel consumption (10^{-12})	0.0468	0.0621	0.0707
Ecotoxicity ($10^3\,m^3$)	2.78	5.52	5.81
Greenhouse effect gas (kg CO_2)	55.2	58.6	53.4
Acidic gas (kg SO_2)	0.252	0.321	0.306
Air pollution (kg C_2H_2)	0.0297	0.0453	0.0460
Toxicity (kg)	0.318	0.429	0.411
Primary energy (MJ)	461	592	643
Waste (kg)	36.3	59.6	58.8

Note: Transport distance from factory to construction site is assumed to be the same.

of hollow core precast concrete contributes more to the reduction of energy consumption than half-precast concrete or on-site placement concrete as shown in Table 5.9. Likewise, the use of hollow core precast concrete reduces the amounts of generated harmful substances and waste. In addition, the use of precast concrete members contributes to the mitigation of energy consumption, and noise and vibration generated at the construction site resulting from reduction of transportation frequency.

Furthermore, as precast concrete members are generally produced in factories, rational methods of energy utilization are feasible. In other words, there are some potential uses, for example, of the waste heat from heat curing which is to accelerate the strength development of concrete as a heat source for the factory itself, and the heat generated by hydrating cement as a heat source for plant forcing culture. As it is essential to have curing equipment for products in the factory, in addition to a heating system for the factory itself, energy consumption for maintaining a warm environment throughout the entire factory may become very high in cold regions.

In warm regions, meanwhile, it is possible to eliminate heat curing processes including steam curing through effective use of the heat generated from the hydration of cement when using insulated formwork.

As precast concrete members are produced in a factory, they are under better quality control than those produced by on-site placement, and the quality of their surface layer is improved while inconsistencies in the covering depth are reduced. Therefore precast concrete has greater resistance concerning the corrosion of reinforcing bars compared with that of on-site placement concrete, thereby the service life of concrete buildings is longer.

5.3.6 Reduction of Reinforcement

A greater number of reinforcing bars is required for high-rise reinforced concrete buildings constructed in seismic regions, and they become very dense, particularly in column-to-beam joints, which in some cases creates difficulty in pouring concrete. Also, bending work during the reinforcing bar process has become difficult due to requirements concerning higher-strength and thicker bars. To solve these problems, the mechanical anchorage method has been introduced, for example by attaching fixed hardware to the ends of the bar or to form its ends in a T-shape (with 2.5 times the diameter of the bar).

Figure 5.26 shows examples of mechanical anchorage bars, and Figures 5.27 and 5.28 show examples of the mechanical anchorage installation. This method makes it possible to reduce the anchorage bars of the main reinforcement and shear reinforcement in places where bars are arranged densely.

FIGURE 5.26
Application of mechanical anchorage bars.

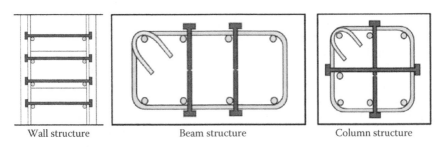

| Wall structure | Beam structure | Column structure |

FIGURE 5.27
Examples of mechanical anchorage bar installations.

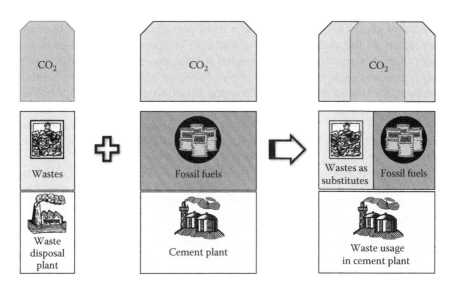

FIGURE 5.28
Reduction of CO_2 emissions by use of wastes.

For example, in the case of an eighteen-story apartment building (total floor area of 9,837 m²) constructed in Japan, the amount of reinforcing bars can be reduced by approximately 32 tonnes (approximately 2% of 1,554 tonnes of total bars). In the case of a box-type caisson and substructure of a viaduct (pier height: 5.0 m), bars can be reduced by approximately 4.8 tonnes (approximately 10% of 47.0 tonnes of total bars) and approximately 3.7 tonnes (approximately 6% of 64.0 tonnes of total bars), respectively.

In addition, having the same level or more anchor strength as the conventional anchorage bars, the method can contribute to the improvement of constructability including simplification of the column-beam joint areas and elimination of complex anchorage bars. Thus the mechanical anchorage method, although increasing cost due to the extra expense for processing the bar heads, reduces not only the amount of reinforcing bars, but also energy for assembling as well as the construction period.

5.3.7 Utilization of Renewable Energy and Low-Energy Machinery

Most of the energy we utilize is derived from exhaustible resources represented by fossil fuels (oil, natural gas, oil sand, and methane hydrate) and underground resources such as uranium (nuclear power generation). In recent years however, the use of non-exhaustible energy has been increasing as a potential energy source with newly found advantages, such as measures to counter the finite nature of exhaustible resources and mitigation measures for global warming.

As non-exhaustible energy is taken from phenomena that recur over and over again in the natural environment and is spontaneously regenerated faster than it is consumed, it allows for sustainable use in the future. With energy sources such as hydraulic power, biomass, solar, wind, and geothermal heat, its use is attracting increasing attention also in concrete-related areas. Furthermore, hybrid type construction machinery has been developed which realizes both low fuel consumption and low carbon dioxide emissions by not relying on fossil fuel alone, but combining several power sources while making good use of each of their advantages.

Biomass is organic matter and generates CO_2 when used as energy through its combustion, but at the same time plants absorb CO_2 as they grow. Thus seen as a whole, it offsets the amount of CO_2 released, having characteristics of "carbon neutrality." By utilizing this biomass as an alternative to fossil fuels, the amount of CO_2 emissions can be reduced.

The biomass energy source is classified into waste and plant matter (cultivated crops) based on its raw material nature. Waste-based biomass uses the electricity and heat generated by the combustion of industrial waste (black liquor and waste woodchips from paper manufacturers, waste and by-products such as rice husks and cattle feces from agriculture, forestry and livestock industries and domestic waste such as rubbish and cooking oil).

Plant-based biomass uses plants such as sugar cane and rapeseed by converting them to alcohol fuel and the like. In the United States, sources for biomass energy consist of plant waste (36%), black liquor (30%), wood waste (25%), and biogas (6%), accounting for approximately 3% of its gross primary energy supply. In EU countries on the other hand, the sources are predominantly firewoods in the conventional forms, while black liquor from paper mills, waste wood from woodwork shops, and municipal waste are co-combusted with fossil fuels for power generation. Their biomass energy accounts for approximately 3.3% of the gross primary energy supply in the EU region.

Such efforts concerning the use of biomass energy are also actively carried out in the cement manufacturing industry. As the use of waste-based biomass such as woodchips and bone-meal feed, and chemical-based industrial waste such as waste plastic and tires expands, it becomes a factor regarding the increase of electric power consumption in cement factories due to incidental pretreatment and in-plant transportation.

However, even allowing for such adverse factors, the use of waste contributes to the reduction of overall energy consumption. When the waste-based biomass and chemical-based industrial waste are not effectively utilized by the cement industry, there is no other choice but to incinerate or dispose of the materials in landfills, and carbon contained in the waste ultimately becomes CO_2 or CH_4 which has an intense greenhouse effect.

However, when they are used effectively in the cement industry, in addition to lower amounts of fossil fuels replaced by the waste-based fuels, CO_2 emissions will be reduced based on the total emissions generated by cement factories and simple incineration facilities, as shown in Figure 5.29.

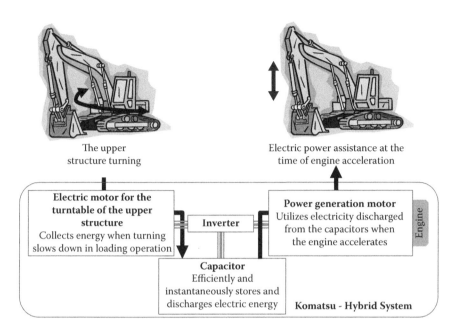

FIGURE 5.29
Hybrid hydraulic excavator mechanism.

The industry has also been trying to reduce the use of natural raw materials through effective utilization of coal ash, sewage sludge, and municipal rubbish. Most of the unused waste materials are disposed of in landfills with or without incineration. Such efforts made by the cement industry to effectively reuse these wastes as raw materials for cement and energy replacement contribute to reducing new construction for final disposal sites and the energy required for disposal activities. The reuse of waste materials with a high CaO content also contributes to the reduction of CO_2, by reducing its source material.

There is also an attempt to put electric energy generated by solar and wind power to use for the maintenance of concrete structures. For example, an electrolytic protection system for reinforcing bars in concrete and a health monitoring system for concrete structures have been developed through the use of electricity generated by natural energy such as solar or wind power.

An example of hybrid-type construction machinery is the hybrid hydraulic excavator. As shown in Figure 5.30, the excavator has a mechanism that reuses the energy generated when it slows down to rotate its body; it is converted to electric energy and stored in a capacitor (condenser), as auxiliary power for accelerating its diesel engine through a power generation motor. This excavator realizes reduction of fuel consumption of approximately 25% compared with ordinary excavators. Furthermore another type of hybrid hydraulic excavator has also been developed, which has the function to

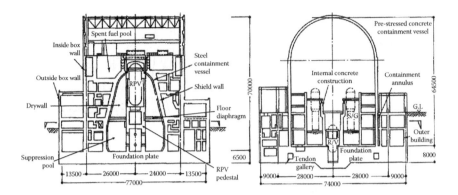

FIGURE 5.30
Nuclear reactor buildings.

supply the power stored as excess energy of a diesel engine in a battery to the engine when it is under heavy load. This type reduces fuel consumption by approximately 50%.

Other developments include electric tractors with 20% lower fuel consumption and 10% lower life-time running costs, hybrid wheel loaders with 10% lower fuel consumption which was realized by installing an automatic idling stop function or by controlling excess engine revolutions of the diesel engine with the assistance of an electric motor.

5.3.8 Protection of Living and Natural Environments

Some concrete structures serve to prevent man-made hazardous substances from leaking or flowing out into the atmosphere, hydrosphere, geosphere, and other natural environments and posing dangers to the natural environment and humankind. Others are built for the purpose of defending the natural environment and humankind from natural disasters.

Examples of the former include the buildings of nuclear power stations, particularly the shielding walls (for boiling water type light-water reactors) and reactor containment vessels (for pressurized water type light-water reactors) in the nuclear reactor building shown in Figure 5.30, are constructed with thick concrete which has a radiation shielding property. Concrete shields gamma and neutron radiation, and its gamma radiation shielding property is more or less proportional to the density and thickness of the shielding body.

Meanwhile in order to shield neutron radiation, it is necessary that concrete contains relatively high atomic number elements that can moderate fast neutrons, light elements like hydrogen that can change intermediate neutrons to thermal neutrons, and materials that can absorb such thermal neutrons. Concrete is a composite of hydrogen, other light elements, and high atomic number elements, and thus is used as a shielding material not only for gamma radiation but also neutron radiation.

Other examples include pipes and box culverts for a sewage system which comprises the social infrastructure while transferring soiled materials such as domestic waste water and industrial effluents out of the urban area, improving public sanitation and preventing the pollution of rivers and lakes and dewatering ponds that serve as reservoirs for waste water and effluent from mines, refineries, and chemical plants.

Most of the structures that protect our living and natural environments against natural disasters are made with concrete, including snow prevention embankments to stop avalanches, river levees against flooding, breakwaters and storm surge barriers to mitigate damage from surges and tsunamis, flood-control dams, and sand control dams to prevent damage from landslides.

Thus concrete structures greatly contribute to the creation of a sustainable society, while protecting human lives and living space from hazardous substances and natural calamities.

References

Architectural Institute of Japan. 2008. Recommendations for environmentally conscious practice of reinforced concrete buildings.

Betong och Miljo. 1999. Betongforum c/o Cementa. Stockholm.

CEMBUREAU. 2010. *Activity Report*. European Cement Association, Brussels.

Constantz, B., Holland, T.T., and Clodic, L. 2010. Sequestering CO_2 in the built environment. *Proceedings of International Conference on Future Concrete*, Doha, pp. 104–111.

De Saulles, T. 2006. Utilisation of Thermal Mass in Non-Residential Buildings. The Concrete Centre.

Fluitman, A. and de Lange, V. P. A. 1996. CREM Report No.95.107, Comparison of the Environmental Effects of Three Concrete Story Floors. Amsterdam.

Gartner, E. and Quillin, K. 2007. Low-CO_2 cements based on calcium sulfoaluminates. *Proceedings of International Conference on Sustainability in the Cement and Concrete Industry*, Lillehammer, Norway, pp. 95–105.

Hashimoto, I. and Watanabe, T. 1999. Clinker burning in the fluidized bed: an innovative technology. *ZKG International*, Vol. 52, No. 1, pp. 1–19.

Humphreys, K. and Mahasenan, M. 2002. Toward a sustainable cement industry. Substudy 8: Climate change. World Business Council for Sustainable Development.

IEA. 2008a. *Energy Statistics: Electricity for the World*. International Energy Agency, Paris. http://www.iea.org/stats/electricitydata.asp?COUNTRY_CODE=29

IEA. 2008b. *Energy Technologies Perspective*. International Energy Agency, Paris.

IEA. 2011. *Energy Statistics: Electricity for the World*. International Energy Agency, Paris.

IPCC. 2007. *Fourth Periodic Assessment Report*. Intergovernment Panel on Climate Change.

Li, G. S., , Walenta, G., and Gartner, E. M. 2007. Formation and hydration of low-CO_2 cements based on belite, calcium sulfoaluminate, and calcium aluminoferrite. *Proceedings of 12th International Congress on the Chemistry of Cement*, Montreal, Paper TH3-15.

Mehta, P. K. 2009. Global concrete industry sustainability: tools for moving forward to cut carbon emissions. *Concrete International*, ACI, Vol. 321, No. 2, pp. 45–48.

Nippon Slag Association. 2011. http://www.slg.jp/slag/index.html

Novacem. 2001. http://novacem.com/technology/novacem-technology/

Sakai, K. 2009. Environmental management of infrastructures: concrete structures. *Concrete Journal*, Vol. 47, No. 2, pp. 17–23.

Tatebayashi, H., Okamoto, M., Nishimura, M. et al. 2000. Experimental studies on production of high quality recycle aggregate from demolished concrete. *Proceedings of Japan Concrete Institute*, Vol. 22, No. 2, pp. 1099–1104.

The Concrete Centre. 2008. Thermal Mass for Housing.

Tongbo, S., Zhaijun, W., Jing, W. et al. 2007. Study on the performance of high belite cement and concrete. *Proceedings of International Conference on Sustainability in the Cement and Concrete Industry*, Lillehammer Norway, pp. 257–262.

Uchiyama, S., Kuroda, Y., and Hashida H. 2003. Deep mixing stabilization using concrete power from recycled aggregate production by heating and rubbing. *Foundation for Engineering & Equipment*, pp. 62–65.

Wang, J., Baco, I., Morin, V. et al. 2010. Hydration mechanism of cements based on low-CO_2 clinkers containing belite, yeelimite and calcium alumino-ferrite. *Proceedings of 7th International Symposium on Cement and Concrete*, Jinan, pp. 127–136.

WBCSD Cement Sustainability Initiative. 2009. *Recycling Concrete*. World Business Council for Sustainable Development.

World Steel Association. 2010. http://www.worldsteel.org/?action=stats&type=steel&period=latest&month=13&year=2010

Yanagibashi, K., Yonezawa, T., Kamiyama, T. et al. 1999. Study on high quality recycle aggregates. *Proceedings of Japan Concrete Institute*, Vol. 21, pp. 205–210.

Yoda, K., Shintani, A., Takahashi, I. et al. 2004. Quality of recycles of coarse and fine aggregates produced by mechanical grinding. *Proceedings of Japan Concrete Institute*, Vol. 26, No. 1, pp. 1627–1632.

6

Sustainable Concrete Technologies: Case Studies

6.1 CO$_2$ Negative Concrete in Japan

As approximately 30% of CO$_2$ emissions are generated by the electric power industry in Japan, improvement in thermal efficiency at thermal power plants and increased use of renewable energy such as hydro, solar, and wind power have been promoted as measures to reduce CO$_2$ emissions. In the concrete sector also, efforts have been underway for CO$_2$ reduction.

The *Recommendation of Environmental Performance Verification for Concrete Structures* published by the Japan Society of Civil Engineers (JSCE, 2000) introduces, as a means of CO$_2$ reduction in concrete structures, the optimization of structural forms and building methods and effective use of by-products, showing related examples. In terms of the use of by-products, fly ash and blast furnace slag are commonly known. However, their use as blended cement remains at little more than 20% of the total due to some problems, for example, the initial strength of fly ash decreases and blast furnace slag has low concrete crack resistance.

Chugoku Electric Power Co., Inc., Kajima Corporation, and Denki Kagaku Kogyo Kabushiki Kaisha have co-developed CO$_2$ negative concrete that realizes "zero or less" CO$_2$ emissions from concrete by forcing concrete to absorb the CO$_2$ emitted from thermal power plants. CO$_2$ negative concrete also uses the coal ash emitted from thermal power plants as a cement substitute. In other words, the development resulted in using both CO$_2$ and coal ash—"nuisance" by-products from thermal power plants—as its materials.

CO$_2$ negative concrete consists of conventional concrete materials (water, cement, and aggregate), coal ash, and a special admixture that reacts with CO$_2$. By curing the concrete with such ingredients under CO$_2$ exposure for about two weeks, it will harden through the processes of carbonation of CO$_2$ and a special admixture in addition to the hydration of water and cement. Table 6.1 shows the differences between ordinary concrete and CO$_2$ negative concrete.

TABLE 6.1

Comparison of Ordinary Concrete and CO_2 Negative Concrete

	Ordinary Concrete	CO_2 Negative Concrete
Materials	Water + cement + aggregate	Water + cement + aggregate + special admixture + coal ash
Curing	Water curing or air curing	Curing by CO_2 in gases emitted from thermal power station
Hardening reaction	Hydration reaction of water and cement	Carbonation reaction of CO_2 and special admixture plus hydration reaction of water and cement

The special admixture is dicalcium silicate γ phase, or γ-$2CaO·SiO_2$ (hereafter shown as γ-C_2S). Among several production methods of this admixture, one that uses raw materials that generate less CO_2 was selected so as to lower CO_2 emissions during its production, to 20% of those emitted during the production of ordinary cement.

As γ-C_2S is nonreactive with water, it does not show any strength development when used in ordinary concrete, but shows strength development equivalent to or higher than cement through its reaction with CO_2.

Average concrete structures also have the so-called carbonation phenomenon by reacting to CO_2 in the atmosphere. However, the reaction speed is very slow, reaching a depth of only a few millimeters to centimeters of a structure's surface over several decades. On the other hand, CO_2 negative concrete, as it can absorb a large amount of CO_2 subject to proper adjustment of temperature and humidity, achieves a substantial reduction of CO_2 emissions, even when taking the amount arising from concrete into consideration.

As shown in Figure 6.1, the process involves transport of the off-gas generated from fuel combustion at a coal-fired power plant to a concrete curing chamber to allow the CO_2 contained in the off-gas to be absorbed by concrete. The off-gas is returned to the system from the chamber and discharged from the chimney. CO_2 contained in the off-gas is 15 to 20% when entering the curing chamber, and 11 to 13% when leaving it after being cured. In order to utilize CO_2 from the off-gas from thermal power plants, two weeks of the curing process are required, the reduction of which remains as a future task.

With respect to the minute amounts of NOx and SOx contained in the off-gas, tests have been conducted to ascertain their impact on the performance of concrete. It was revealed that while they have hardly any influence on the strength and porosity of concrete, they do, however, inhibit the speed of concrete carbonation. Therefore the gas is used after eliminating NOx and SOx.

Figure 6.2 shows a comparison of CO_2 emissions from ordinary concrete and CO_2 negative concrete used for boundary and paving blocks. As the

FIGURE 6.1
Production of CO_2 negative concrete.

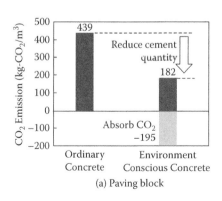

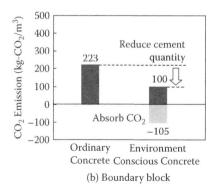

FIGURE 6.2
Comparison of carbon dioxide emissions from ordinary concrete and CO_2 negative concrete.

FIGURE 6.3
Application of CO_2 negative concrete to paving blocks.

figure demonstrates, CO_2 emissions during the processes from the production of cement as a concrete material, curing of concrete, and finalization of the product are zero or less. The compressive strength of boundary blocks and bending strength of paving blocks both satisfy 18 N/mm² and 5 N/mm², respectively, as required by the Japanese Industrial Standards (JIS). The mix proportion of the CO_2 negative concrete introduced here has not been publicized, but values of 159.3 kg CO_2/t and those based on the JSCE guidelines (2006) were used for the unit-based CO_2 data of γ-C_2S and the rest of the ingredients, respectively. Figure 6.3 shows the application of CO_2 negative concrete to paving blocks.

6.2 Low Carbon High-Flowable Concrete

It is possible to reduce a considerable amount of CO_2 emissions compared with conventional concrete, by replacing cement that generates large carbon output with SCMs such as ground granulated blast furnace slag. Obayashi Corporation developed low carbon high-flowable concrete, while replacing ordinary Portland cement (C) with ground granulated blast furnace slag (BS), fly ash (FA), and zirconia origin silica fume (SF) as binder materials.

Studies were made by setting the different cement blending ratios from 100% to 10%, and substituting the remaining part with SCM(s), as shown in Table 6.2. Figure 6.4 shows the compressive strength of each mix proportion. It reveals that by setting the proper water-binder ratios, even when the cement blending ratio is less than 30%, the specified performances are obtained. Figure 6.5 shows that such a blending ratio achieves CO_2

TABLE 6.2

Cement: SCM Ratios

	Blend Ratio (%)			
Cement	Blast Furnace Slag	Fine Aggregate	Silica Fume	Water (kg/m³)
100	0	0	0	160
50	50	0	0	152
25	75	0	0	149
25	65	10	0	144
25	55	20	0	137
15	85	0	0	147
15	75	10	0	144
15	65	20	0	137
15	55	30	0	132
15	65	17.5	2.5	135
15	65	15	5	135
10	90	0	0	147

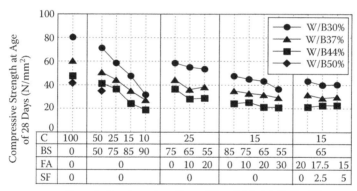

FIGURE 6.4

Compressive strengths of different cement blends.

reduction up to a level of 50kg/m³. Table 6.3 shows the inventory data used for the calculation of CO_2 emissions. Figures 6.6 through 6.8 show the results of drying shrinkage, accelerated carbonation testing, and freezing and thawing tests, respectively.

Based on these successful results, low carbon high-flowable concrete has been applied to a concrete fence wall (four-component system), seismic concrete block wall (four-component system), and temporary pedestrian path (three-component system) as shown in Figures 6.9, 6.10, and 6.11.

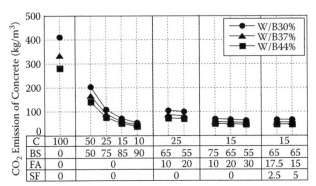

FIGURE 6.5
Blending ratios to achieve CO_2 reductions up to 50 kg/m^3.

TABLE 6.3

CO_2 Emission Data for Materials

Material	CO_2 Emissions (kg-CO_2/t)
Portland cement (C)	757.9
Blast furnace slag (BS)	24.1
Fly ash (FA)	17.9
Silica fume (SF)	17.9
Water	0
Fine aggregate	3.5
Coarse aggregate	2.8
AE high-range water-reducing admixture	200

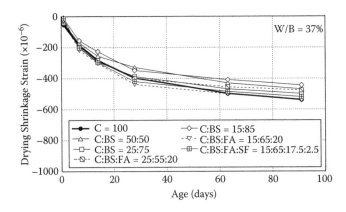

FIGURE 6.6
Results of drying shrinkage test.

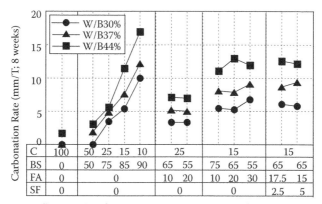

FIGURE 6.7

Results of accelerated carbonation test.

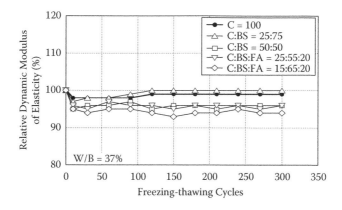

FIGURE 6.8

Results of freezing and thawing test.

6.3 High-Performance Shotcrete

The technology of using fly ash in shotcrete in order to reduce environmental impacts has been put into practice (fib, 2009). Coal ash generated from thermal power plants is classified as HFA to produce fly ash for this purpose and has the chemical and physical properties listed in Table 6.4. Table 6.5 shows three examples of shotcrete mixture proportions. The basic concept of HFA shotcrete is the replacement of fine aggregates with fly ash. Figure 6.12 shows the compressive strength for each mixture proportion, and reveals that HFA shotcrete demonstrates remarkably high strength.

FIGURE 6.9
Concrete fence wall.

FIGURE 6.10
Seismic concrete block wall.

Using HFA shotcrete reduces environmental impacts in the following three ways:

1. Reduction in the use of natural resources and extension of the service lives of waste disposal facilities
2. Reduction in emissions of CO_2, a greenhouse gas
3. Reduction in emissions of CO_2 attributed to reduced rebound rate

In the Shikoku region of Japan, approximately 12,000 tonnes of HFA was used in 2006, thus avoiding the use of the equivalent amount of natural fine aggregates and extending the service lives of facilities that would otherwise have

FIGURE 6.11
Temporary pedestrian path.

TABLE 6.4

Chemical and Physical Properties of HFA

	Chemical Property			Physical Property	
	SiO_2 (%)	Wet Percentage (%)	Loss on Ignition (%)	Density (g/cm³)	Specific Surface (cm²/g)
HFA	57.0	0.09	1.5	2.40	5570

TABLE 6.5

Shotcrete Mixture Proportions

Mixture Proportion Type	Water-Cement Ratio, W/C (%)	Sand-Coarse Aggregate ratio, s/a (%)	Unit Quantity (kg/m³)					
			Water	Cement	HFA	FA	CA	Dust Suppressant
Conventional mix 1	61.4		221	360	–	1038	707	–
Conventional mix 2	61.4	60.3	221	360	–	1038	707	0.36
HFA mix	58.9		212	360	100	943	718	–

Note: Maximum size of coarse aggregate (CA) 15 mm; slump 8 ± 2; set accelerating agent.

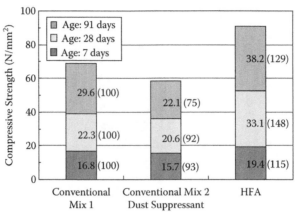

<pre>*Figures in parentheses indicate percentage, with the value for
conventional mix 1 taken as 100%.</pre>

FIGURE 6.12
Comparison of compressive strengths of shotcrete mixtures.

had to dispose of the equivalent amount of coal ash. At the same time, CO_2 emissions associated with the production of fine aggregates were reduced by approximately 50 tonnes. HFA shotcrete also has a considerably improved rebound rate—the dust level and rebound rate are shown in Figure 6.13, which reveals that the values are approximately half those of shotcrete with conventional mixture proportions.

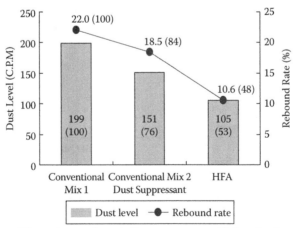

<pre>*Figures in parentheses indicate percentage, with the value for
conventional mix 1 taken as 100%.</pre>

FIGURE 6.13
Dust level and rebound rate comparison.

This considerable reduction in the dust level leads to remarkable improvement of the working environment. Moreover, the halving of the rebound rate results in a corresponding halving of the amount of shotcrete used. It is estimated that using shotcrete with 12,000 tonnes of HFA is equivalent to a CO_2 emission cut of approximately 4,980 tonnes. To date, HFA shotcrete has been applied to more than fifty tunnels in Japan.

In this case, the reduction in CO_2 emissions in terms of absolute value was not so significant. However, it can be said that the use of fly ash to reduce environmental impacts was shown to be an excellent technology.

6.4 Closed-Loop Concrete System on Construction Site

Most construction waste is concrete rubble and the stock of concrete level is large. Most concrete waste had been reused for road sub-bases. However, concrete waste was predicted to exceed the demand for sub-base rubble in near future, resulting in over-fill of disposal sites. Therefore recycling concrete rubble is an urgent issue in the concrete field.

Shimizu Corporation developed the closed-loop concrete system (Kuroda et al., 2002) and applied it to a construction site. The aims of this system are to maintain the high quality of aggregate and to completely use concrete waste. Figure 6.14 shows the job site in the construction process. This system was adopted to contribute to environmental protection of the community in the rebuilding of a warehouse. Figure 6.15 shows the exterior view of the

FIGURE 6.14
Closed-loop concrete system at job site.

FIGURE 6.15
Exterior view of warehouse before demolition.

warehouse before demolition and Figure 6.16 shows completion of the project. This project's demonstration of making high quality recycled aggregate triggered the formulation of JIS 5021 covering recycled aggregate for concrete class H.

Figure 6.17 shows the recycling flow of the closed-loop concrete system. Concrete lumps from a demolished building, which had been confirmed to be recyclable beforehand, are source-separated during demolition and

FIGURE 6.16
Completed project.

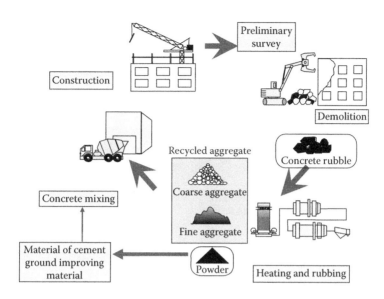

FIGURE 6.17
Recycling flow of closed-loop concrete system.

processed to produce recycled coarse aggregate and recycled fine aggregate having qualities comparable to those of the original aggregates. These are used for recycled aggregate concrete for structural use, while the fine powder derived from cement paste and separated from the aggregates is used for cement materials and ground improving materials.

Figure 6.18 shows recycled coarse aggregate, fine aggregate, and fine powder. Since these recycled aggregates retain their original qualities, they can be reused for the same uses, thus establishing a closed recycling loop.

This system adopted "heating and rubbing" as this was the only method that could produce both coarse and fine recycled aggregates with high qualities. This is a technology to selectively remove cement hydrate from the aggregate, and this becomes possible by rubbing concrete rubble after the

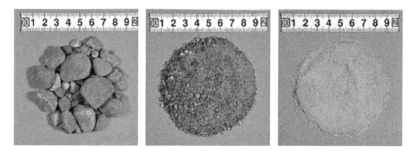

FIGURE 6.18
Recycled aggregates.

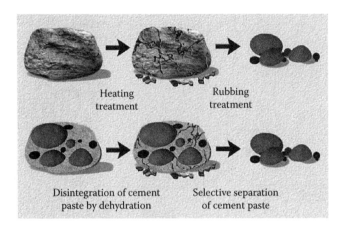

FIGURE 6.19
Heating and rubbing method.

cement hydrate has been made fragile by heating to about 300°C. Figure 6.19 shows the concept of the heating and rubbing method.

Prior to construction, an investigation was conducted to check whether the aggregate to be recovered from demolished concrete was suitable for recycling because no record on the qualities of aggregate was found in the literature. Thus it was decided to conduct an investigation on the exterior appearance and on the core samples obtained by test drilling to evaluate recycling possibilities.

Accordingly, the appearance revealed no deterioration due to chloride attack or alkali-aggregate reaction. The chloride content of concrete was lower than the specified detection limits, and the water absorption and oven-dry density of the original aggregate comfortably complied with the JIS requirements for natural aggregate. All aggregates were proven innocuous in regard to alkali–silica reactivity. Based on these results, the concrete was judged to have a high possibility of containing suitable aggregate, and it is recyclable.

The aggregate recycling plant using the heating and rubbing method, Figure 6.20, is a mobile type allowing disassembly into approximately cubic frames with 3 m sides, transportation, and reassembly. The aggregate recycling plant was designed to have a target maximum treatment capacity of 9 t/hour.

All qualities of recycled aggregate satisfied the JIS requirements and were high even in comparison with the original aggregate. Concrete rubble of about 34,500 tonnes (about 31,000 tonnes on a dry base) was treated in the aggregate recycling plant during the construction period. The collection rate of the recycled coarse aggregate was about 35%, and the collection rate of the recycled fine aggregate was about 21%.

Table 6.6 gives the mixture proportions of concrete. Three types of recycled aggregate concrete were used in the project: one with recycled fine aggregate and recycled coarse aggregate (RR); one with recycled fine aggregate

FIGURE 6.20
Mobile aggregate recycling unit.

and normal coarse aggregate (RN); and one with normal fine aggregate and recycled coarse aggregate (NR). Normal aggregate concrete (NN) was also produced for comparison by the on-site plant.

The slump, air content, and compressive strength also satisfied the control values. Figure 6.21 shows typical compression test results of standard-cured

TABLE 6.6

Mix Proportions

| Mark | Nominal Strength | W/C % | Unit Weight (kg/m³) | | | | |
			C	W	S	G	Ad. (C × %)
RR	27	51.7	323	167	792	1011	0.250
	30	48.4	347	168	769	1011	0.250
	33	45.5	374	170	743	1011	0.250
RN	27	54.4	320	174	829	972	0.250
	30	50.4	345	174	808	972	0.250
	33	47.0	372	175	783	972	0.250
NR	27	51.7	323	167	722	1080	0.250
	30	48.4	347	168	699	1080	0.250
	33	45.5	374	170	673	1080	0.250
NN	27	54.4	320	174	810	988	0.250
	30	50.4	345	174	789	988	0.375
	33	47.0	372	175	764	988	0.375

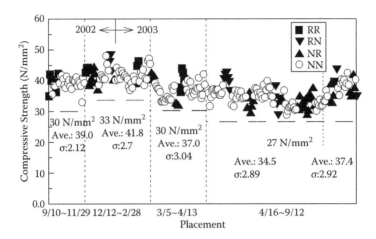

FIGURE 6.21
Compression test results.

concretes at the control age of 28 days. The standard deviation of the compressive strength ranged between 2 and 3 N/mm², showing relatively small variation in quality similar to NN.

In order to accomplish the closed-loop concrete system, usage of the byproduct powder discharged from recycled aggregate is very important. Table 6.7 shows an example of the chemical composition and physical properties of the powder. The powder has a large specific surface area and includes a cement component.

Table 6.8 shows an example of the uniaxial compressive strength of hardened powder paste and cement paste. The slight strength development of powder paste was caused by a little hydraulicity of the powder itself. Therefore, the powder can be used for ground improvement as a partial substitute for cement, when high-strength development is not required. The powder was used for shallow ground improvement and deep ground improvement.

TABLE 6.7

Chemical Composition and Physical Properties of Powder

	Density g/cm³	Blain cm²/g	Chemical Composition (%)								
			Ig. Loss	SiO_2	Al_2O_3	Fe_2O_3	CaO	MgO	SO_3	Cl	R_2O
Sample 1	2.48	5,510	9.6	50.8	10.1	2.7	23.5	1.3	0.8	0.02	1.94
Sample 2	2.47	6,280	9.8	41.9	12.0	6.5	23.2	2.1	0.9	0.06	2.08
OPC	3.15	3,450	1.5	21.2	5.2	2.8	64.2	1.5	2.0	0.01	0.63
BB	3.03	3,970	1.6	25.6	8.5	1.8	54.7	3.6	2.0	0.01	0.50

OPC = Ordinary Portland cement. BB = Blast-furnace slag cement.

TABLE 6.8

Uniaxial Compressive Strengths of Hardened Power Paste and Cement Paste (N/mm²)

	Powder Paste (Samples Used)			Cement Paste (Blast-Furnace Slag Cement)		
Age (Days)	W/P = 0.50	W/P = 0.75	W/P = 1.00	W/C = 0.50	W/C = 0.75	W/C = 1.00
3	0.11	0.07	0.03	–	–	–
7	0.49	0.21	0.10	23.6	9.85	6.21
28	1.70	0.60	0.32	50.1	27.8	16.4

A life cycle assessment was performed assuming the case of production of road sub-bases by transporting concrete rubble to the intermediate processing facilities (usual case for comparison) and the case of applying the closed-loop concrete system at a construction site. The impacts were analyzed for each stage of P1 (road sub-base production), P2 (recycled aggregate recovery), P3 (concrete mixing), and P4 (ground improvement). The amounts of CO_2 emissions for each process were calculated. Table 6.9 shows the inventory data on CO_2. The calculation results are shown in Figure 6.22.

The CO_2 emission of recycled aggregate recovery was 71.2 kg-CO_2/t, which was five times or more than the environmental impact of road sub-base production. However, the emission amount was reduced by about 1,000 t-CO_2 because the CO_2 emission in P3 was reduced due to a decrease in transportation, and the CO_2 emission in P4 was reduced due to a decrease in the amount of cement.

Thus, the closed-loop concrete system provides a means of creating a recycling-oriented society while contributing to environmental protection of the community. Nevertheless, technical problems remain regarding energy consumption for heating and rubbing and the use of by-product powder when

TABLE 6.9

CO_2 Inventory Data

Inventory		CO_2 Emission (kg-CO_2/t)
Transportation (truck), km		0.314
Road sub-base production		13.578
Recycled aggregate production	Crushing	12.592
	Heating and rubbing	58.556
Concrete mixing (ready-mixed concrete plant)		3.192
Concrete mixing (site plant)		2.714
Materials	Crushed stone	11.033
	Mountain sand	12.355
	Ordinary Portland cement	800.116
	Cement soil stabilizer	502.000

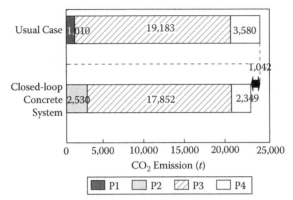

(P1: Road sub-bases production, P2: Recycled aggregate recovering,
P3: Concrete mixing and delivery, P4: Ground improvement)

FIGURE 6.22
Results of life cycle assessment of CO_2 emissions.

a large amount of waste concrete is processed. Waste heat management and the use of fine powder for cement materials would be the best solutions to these problems.

6.5 Concrete Pavement

The performance differences between concrete and asphalt pavements have been the subjects of discussion. Although asphalt pavement provides a smooth surface and good ride quality, age-related deterioration and rutting occur in a relatively short time, requiring re-surfacing after ten years. On the other hand, concrete pavement can be made sufficiently resistant to fatigue resulting from repeated loads. If subjected to appropriate design and construction, it is not likely to develop fatigue cracks within the period specified by its design, and even when such cracks do occur, it is possible to further extend its structural life by proper repair.

Another advantage is that unlike asphalt, concrete pavement does not produce flow rutting as it does not develop plastic deformation. It also has excellent abrasion resistance. Based on these factors, the structural durability of concrete pavement can be secured for a period of thirty years or more. The Japan Cement Association has compiled the results of its wide-ranging study concerning the superiority of concrete pavement (JCA, 2009). Such results are introduced in this book.

As a result of its survey, it was found that 25% of Japanese concrete pavements have been in service for more than thirty years with their original

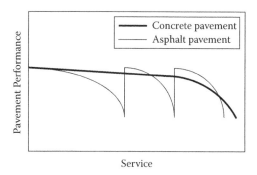

FIGURE 6.23
Changes in pavement performance.

concrete surfaces, and that there are many places that were overlaid with asphalt mixtures before structural fractures occurred, creating composite surfaces. Some have been in use for more than sixty years.

Figure 6.23 shows a conceptual diagram of changes in pavement performance. As concrete pavement has high durability regarding both structure and surface performance, repair is hardly necessary. In fact most of pavements currently in service have not undergone any repair since they were constructed, demonstrating that they contribute to a significant reduction of repair work. There are some that have been in service for sixty years and never needed repair. This represents a great advantage concerning the reduction not only of management cost, but also in traffic restriction and rerouting arising from repair work.

During the designed structural life, fatigue cracks do occur; however, they have no adverse effect on the performance of the road surface and sufficient performance can be maintained even after such cracks develop.

In Japan, it is known that the cost of constructing concrete pavements is higher than that of asphalt, which is a major impediment concerning selection of a pavement type. Initial construction cost varies, depending on the pavement structure, possible extended construction period, and the materials, but in the case of a newly constructed standard roadbed and surface layer, the direct construction cost for concrete pavement is 6,200 to 9,600 yen per square meter. Pavement using a dense-graded asphalt mixture costs 5,900 to 8,000 yen per square meter. However, as concrete pavement requires little repair cost as mentioned above, and considering the so-called life cycle cost (LCC), it is less expensive than asphalt pavement.

Figure 6.24 shows an example of LCCs for a concrete pavement in service for a long period and for an asphalt pavement constructed parallel to a concrete pavement on the same route. The costs of construction (roadbed and surface layer) and repair were calculated for the concrete pavement in use based on the current price. This 30-cm thick concrete pavement was constructed in 1986 and at the time of this survey never had any repairs, while the adjacent asphalt pavement section has been repaired twice.

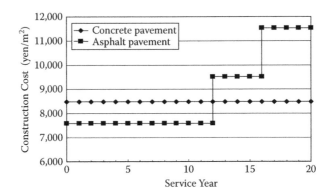

FIGURE 6.24
Comparison of life cycle costs for concrete and asphalt pavements.

This demonstrates that although the initial construction cost is higher in the case of the concrete pavement, if even a single repair involving cutting and overlaying is made to the asphalt pavement to achieve the same life as the concrete pavement, the LCC will be higher in the case of asphalt pavement. According to this calculation, the introduction of concrete pavement can reduce the cost by approximately 25%. A similar calculation was applied to nineteen locations, in seventeen of which the LCC was lower in the case of the concrete pavement than that of asphalt.

In recent years, the heat island phenomenon has been an issue in large urban areas. One of the causes is accumulation of heat in the pavement body. In this respect, concrete pavement absorbs less heat than asphalt pavement, as it has high road surface reflectivity because of its light color, compared with asphalt pavement, which is nearly black. Figure 6.25 shows thermographic images of pavement surfaces which reveal that the asphalt surface represents a temperature of approximately 50°C, while the concrete surface represents approximately 40°C. This is evident in the actual measured temperatures of the road surfaces shown in Figure 6.26, indicating that the concrete surface is cooler by approximately 10°C. Therefore if concrete is used for urban road paving, a certain reduction effect on urban temperature can be expected.

The constituent materials of concrete such as aggregates and cement are basically available locally, thus they can be supplied on a stable basis in terms of both price and volume. As shown in Figure 6.27, as the price of cement is more or less constant, the price of ready-mixed concrete is stable. On the other hand, the price of straight asphalt for paving has soared since 2004 due to the price of crude oil, resulting in raising the cost of asphalt mixture materials also. It is inevitable that the price of asphalt and its mixture materials will continue to be greatly influenced by the price of crude oil.

Another great advantage of concrete pavement is its good visibility in tunnels and at night, as it has high road surface reflectivity because of its light

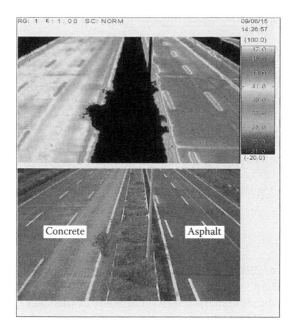

FIGURE 6.25
Thermographic images of pavement surfaces.

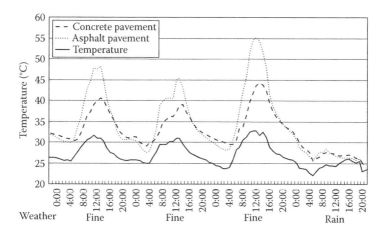

FIGURE 6.26
Road surface temperatures.

color. This provides great benefits for driving and pedestrian safety, while allowing reduction of the energy and cost required for lighting.

Today it is becoming apparent that the type of pavement surface influences the fuel consumption rates of vehicles. It is reported that according to actual measured data by the National Research Council Canada (NRC;

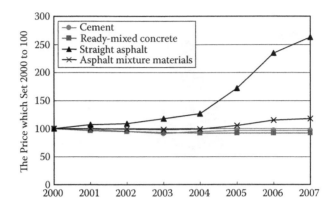

FIGURE 6.27
Cement price stability.

Taylor and Patten, 2006), fuel consumption is reduced by up to 7% in the case of a concrete pavement surface compared with asphalt. It is considered that this effect is attributable mainly to the hardness of the road surface.

In the survey by the Japan Cement Association, no study was conducted concerning life cycle assessment (LCA). As it largely depends on how the specific asphalt consumption is defined, direct evaluation seems difficult. However, it is essential to select the most suitable pavement by carrying out comprehensive assessments of all aspects.

6.6 Ultra-High Strength Fiber Reinforced Concrete

Among all the recent developments of high performance fiber reinforced concrete (HPFRC), studies on ultra-high strength steel fiber reinforced concrete with a compressive strength exceeding $150N/mm^2$ are being carried out actively in many countries. Taisei Corporation is promoting the practical use of such ultra-high strength material, naming it UFC (ultra-high strength fiber reinforced concrete) distinct from ordinary HPFRC. By effectively utilizing its excellent strength properties, the construction of structures with very thin and light members allows reduction of concrete consumption and sizes of substructures, thereby promising a reduction of emissions of CO_2 and other pollutants.

UFC is a new cementitious material consisting of powder materials including cement, silica fume, and silica sand; steel fibers for reinforcement (0.2 mm diameter, 15 mm length, 2% volume ratio); and a special water-reducing agent. UFC has an extremely high average compressive strength of about 200 N/mm^2 and more and shows reduced crack extension and widening from

TABLE 6.10

Comparison of Fundamental Properties of UFC and Conventional Concrete

Item	Unit	UFC[a]	Conventional High Strength Concrete
Compressive strength	N/mm²	180	40
Tensile strength	N/mm²	8.8	2.7
First cracking strength	N/mm²	8.0	1.3
Young's modulus	kN/mm²	50	31
Mass per unit volume	kN/m³	25.5	24.5
Shrinkage strain		50×10^{-6}	130×10^{-6}
Creep coefficient		0.4	2.6
Water permeability coefficient	cm/sec	4×10^{-17}	10^{-10}
Chloride ion diffusion coefficient	cm²/year	0.002	0.7

[a] After standard heat cure (90°C for 48 hr).

the steel fibers. It has a flexural strength of about 35 N/mm² representing an extremely high toughness. Through the application of such reinforcement effect, it is common practice to use no reinforcing bars for UFC structures. Thus the structures are free of restrictions resulting from the positioning of reinforcing bars or the provision of concrete cover. The use of members of 5 to 15 cm thickness has become feasible and would have been unthinkable for structures using conventional concrete.

Table 6.10 shows a comparison of the fundamental properties of UFC after standard steam curing at 90°C for 48 hours and ordinary concrete. As shown in Figure 6.28, UFC contains powders with different particle sizes based on the concept of densest-packing. The unit water content is minimized to the level of a water-to-cement ratio of 22%. As it also contains reactive powders such as silica fume, the water-to-binder ratio is extremely low at about 14%. Therefore, the hardened product is extremely dense, having very high durability with chloride ion ingress of about 10% that of ordinary concrete (W/C: 45%).

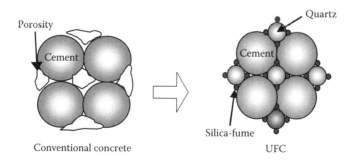

FIGURE 6.28
UFC powder contents.

FIGURE 6.29
Sakata Mirai pedestrian bridge.

In order to introduce the characteristics of UFC structures and to make a comparison with conventional concrete bridges concerning environmental load, the authors have selected the first bridge in Japan to use UFC—the Sakata Mirai pedestrian bridge (Figures 6.29 and 6.30 and Table 6.11). This girder bridge is 50.2 m in length, very long for a single-span structure. Such a long span and a reduction in its dead load were realized through the use of UFC with ultra-high strength in the superstructure. Having an upper deck only 5 cm thick and webs 8 cm thick, the quantity of concrete (UFC) used for this bridge is approximately 20% compared with that of a conventional

FIGURE 6.30
Cross-section of UFC bridge.

TABLE 6.11

Comparison of UFC and Conventional Concrete Bridges

	UFC Bridge	PC Bridge
Structure system	One-span, outer-cable prestressed, unreinforced box-type bridge	Three span, PC simple slab bridge, pre-tensioning system
Length	50.2 m	
Width	2.4 m (full width), 1.6 m (effective width)	
Span	49.35 m	16.7 m
Beam height	0.55 m (at end of bridge) 1.56 m (at center of bridge)	0.6 m
Installation method	Precast block	
Sub-structure	Foundation with steel pipe pile	

bridge, resulting inevitably in reducing its weight to 20% of the weight of a conventional concrete bridge.

When considering the quantity of materials used alone, the application of UFC seems very effective in mitigating environmental load. However, it should be noted that despite its excellent properties concerning strength and durability, it has a high unit cement content due to its unique mix proportions (Table 6.12) and contains steel fibers. Thus, one cannot reach a conclusion based on the fact that it uses a smaller quantity of materials alone. To make a proper comparison of a UFC bridge (Figure 6.31), a prestressed concrete bridge constructed by the conventional method was assumed as an example. It consists of a Portland cement slab deck with three 17-m spans and represents a common design for a bridge of such scale (Figure 6.32).

Setting emissions of CO_2, NOx, SOx, and particulate matters as evaluation indices, and applying the values recommended by the Concrete Committee of the Japan Society of Civil Engineers or the summing-up method, i.e., adding accumulative emissions for each source based on an inter-industry relations table to the inventory data, the amount of environmental load was calculated. The activities subject to calculation were those from the time of the manufacture of materials through the construction of the structures. The construction process and scope of the environmental load assessment are shown in a frame with a broken line (Figure. 6.33).

Emissions arising from the manufacture of materials, manufacture of beams, transport and construction on site were calculated separately based

TABLE 6.12

UFC Mix Proportions (kg/m³)

Water	Cement	Quartz Flour, Silica Sand, etc.	Steel Fiber	Superplasticizer
180	818	1479	157	24

Note: Water includes content in superplasticizer, 10 kg/m³.

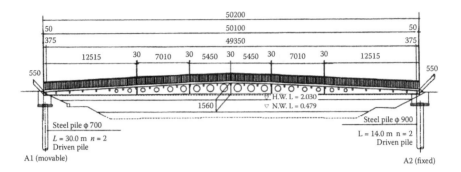

FIGURE 6.31
UFC bridge.

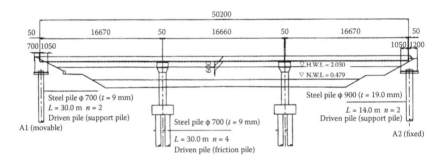

FIGURE 6.32
Prestressed concrete bridge.

on superstructure, substructure, and the temporary structure, respectively. The emissions from manufacture of materials include those from the extraction of raw materials through manufacture of the finished products. The principal material consumption is shown in Table 6.13.

The manufacture of beams parameter includes emissions from manufacturing the main girders and formwork. The main girders of the UFC bridge were manufactured, and those for the PC bridge were assumed to be manufactured in secondary product plants. All girders for the UFC bridge were manufactured using new steel formworks, as the bridge adopted a unique form in harmony with the surrounding landscape, such as an irregular cross section with varying girder height and large circular openings on the webs to create a light-weight appearance. Its symmetrical design made it possible to produce formwork for half of the bridge only. However, its reuse for other bridges and reuse of its steel materials were not taken into account for the calculation. It was assumed that the standard steel formwork was used for the PC bridge and reused 300 times.

The transport factor includes emissions arising from transport of the materials and members, excluding those included in manufacture of

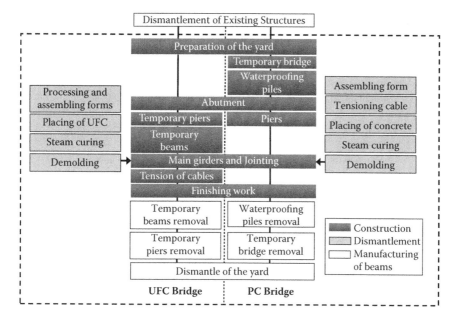

FIGURE 6.33
Environmental load assessment for UFC structure.

TABLE 6.13

Volume of Materials

	UFC Bridge		PC Bridge	
	Superstructure	Substructure	Superstructure	Substructure
UFC (m³)	21	–	–	–
Reinforced concrete (m³)	–	24	51	45
PC steel wire (ton)	3.5	–	2.8	–

materials. Figure 6.34 shows the distribution of construction materials. With respect to emissions from transport required for the re assembling of steel formwork during production of the UFC bridge girders, they were included in manufacture of beams category. Construction on site includes emissions arising from works carried out on site including the use of construction machinery.

6.6.1 CO_2 Emissions

The results of calculations for CO_2 emissions arising from the construction of the UFC and PC bridges were 135 tonnes and 185 tonnes, respectively (Figure. 6.35), showing a possible reduction of 27% in the case of the UFC bridge. With respect to its superstructure where UFC was used, CO_2 emissions were

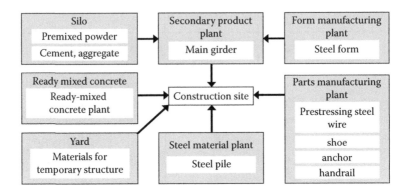

FIGURE 6.34
Distribution of construction materials.

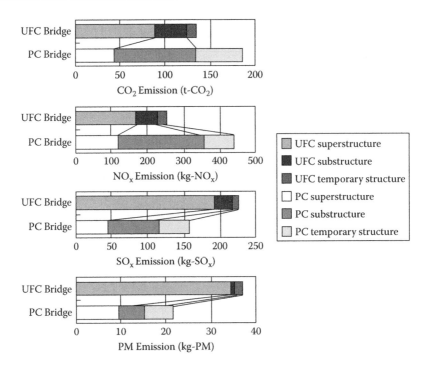

FIGURE 6.35
Emission results from UFC and PC bridges.

greater than those of the PC bridge, while emissions were lower from its substructure and temporary structure as they did not require bridge piers.

With respect to the PC bridge, 78% (72 t) of emissions from the substructure were due to the manufacture of steel pipe piles and 93% (48 t) from the temporary structure were due to the materials for temporary piers and a

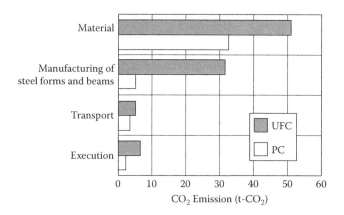

FIGURE 6.36
UFC and PC bridge superstructure emissions.

double-wall cofferdam. On the other hand, CO_2 emissions arising from the manufacture of steel pipe piles used in the abutments of the UFC bridge were 27 t, and a temporary bridge and double-wall cofferdam were not required. Therefore in the UFC bridge, emissions were reduced by 67% and 76% for the substructure and temporary structure, respectively.

The breakdown of emissions from the superstructure using UFC is shown in Figure 6.36. It was revealed that the increase in CO_2 emissions from manufacture of materials was greatly affected by the production of new steel formwork and the fact that its reuse after construction of the UFC bridge was not taken into account. Also, approximately 97% of the 31 t of emissions from manufacture of beams was due to the combustion of heavy fuel oil used for steam curing, which accounted for approximately 22% of the total CO_2 emissions from the UFC bridge.

On the other hand, utilization of the high strength properties of UFC made it possible to lower concrete consumption compared with the PC bridge to less than half. However, the unit cement content was about double that of the PC bridge superstructure (400 kg/m³), so that a significant CO_2 emission reduction effect was not seen as a result of the reduction in quantity of materials used in the superstructure.

6.6.2 NOx Emissions

The results of calculations for NOx emissions from the UFC and PC bridges were 248 kg and 438 kg, respectively (Figure. 6.35), showing a reduction by 43% for the UFC bridge. As in the case of CO_2, the emissions from its superstructure were higher, while a significant reduction was seen in emissions from its substructure and temporary structure by 75% and 73%, respectively.

The increase in NOx emissions from the superstructure of the UFC bridge was greatly affected as in the case of CO_2, by the fact that the reuse of steel formwork and the large consumption of heavy fuel oil for the steam curing were not taken into account. As great amounts of NOx are emitted during the operation of construction machinery, those arising from transport and construction on site accounted for 49% of the total, which was characteristically higher compared with other emitted substances.

6.6.3 SOx Emissions

The results of calculations for SOx emissions from the UFC and PC bridges were 226 kg and 156 kg, respectively (Figure 6.35). In other words, emissions from the former were 45% greater than those from the latter. The emission reduction effect seen in the UFC bridge was as good as in the CO_2 and NOx cases concerning its substructure and the temporary structure, while emissions from its superstructure were very much higher than those from the PC bridge, resulting in the latter exceeding the reduction seen in the former.

The higher emissions from the superstructure and lower emissions from the substructure and temporary structure were due to the same reasons as those in the case of the CO_2 and NOx emissions. However, in the case of SOx emissions, the combustion of heavy fuel oil used for steam curing of the UFC bridge girders accounted for 65% of the total, representing a greater percentage compared with CO_2 emissions.

6.6.4 PM Emissions

The results of calculations for particulate matter (PM) emissions from the UFC and PC bridges were 37 kg and 21 kg, respectively (Figure 6.35). In other words, emissions from the former were 76% greater than those from the latter. As also seen in the case of the SOx emissions, considerable reduction was seen in the substructure and temporary structure, while emissions from its superstructure were very much higher than those from the PC bridge, resulting in the latter exceeding the reduction seen in the former.

Such increase and decrease were due to the same reasons as those for the SOx emissions. In the case of PM emissions, the combustion of heavy fuel oil used for steam curing of the UFC bridge girders accounted for 81% of the total, representing a greater percentage compared with the CO_2 emissions seen in the SOx emissions. As with NOx, PM emissions are considerable during the operation of construction machinery. PM reduction was greater in the case of the UFC bridge in which more power saving during construction was promoted.

A comparison of the environmental loads of the UFC bridge and a conventional Portland concrete bridge revealed that, when comparing them based on their superstructures alone, the UFC bridge was seen to have a higher environmental load, because despite its smaller volume of concrete (UFC) consumption, it produces a large environmental load due to its high unit cement content, the number of times its formwork is reused, and the steam curing. On the other hand, the lighter weight of the UFC bridge attained through the use of UFC contributed to a considerable reduction of the emissions of various substances during construction, and its long-span structure allowed the elimination of bridge piers. Thus, it was seen that when comparing the bridges as a whole combining the superstructure and substructure, the UFC bridge reduced CO_2 and NOx emissions significantly.

Reuse of the formwork was another factor. The UFC bridge has a unique design that harmonizes with the surrounding landscape; the PC bridge has a common and standard form, and therefore the emissions due to manufacture of the formwork between the two bridges was attributed to their designs not the differences in their materials. Furthermore, despite the fact that the greater SOx and PM emissions in the UFC bridge were due mainly to steam curing of its girders one block at a time, this UFC bridge was the first of its kind in Japan and was therefore produced with due care and attention as a prototype. Should this bridge be constructed today, after more than twenty UFC bridges have been constructed, such steam curing would be carried out with several blocks together, using facilities whose curing efficiency has been dramatically improved through the use of thermal insulation.

Thus, SOx and PM emissions arising from present steam curing would be at a level between 50% and 25% of those calculated in the comparison herein. When taking these factors, that is, the number of formwork reuses and steam curing efficiency into consideration, the UFC bridge can be considered as having a low environmental load based on its superstructure alone. When assessed as a whole including the substructure, it is expected that the application of UFC to bridges can greatly reduce environmental load.

6.7 Adaptable Super High Rise Residential Buildings

Super high rise residential buildings are becoming icons for today's big cities. The construction of skyscrapers accelerated following the Great Chicago Fire of 1871. They were initially office buildings with steel-frame construction. However, verticalization of reinforced concrete residential buildings was promoted with the development of high-strength concrete that contributed to increased urban density.

However, conventional reinforced concrete buildings, particularly those in earthquake-prone regions, require many seismic walls to secure resistance to earthquakes, resulting in low flexibility in terms of the use of space. Therefore it was often the case that they were unable to accommodate alterations in floor plans or the renewal of service facilities required with changes of occupants or their lifestyles. As a result, many were demolished and

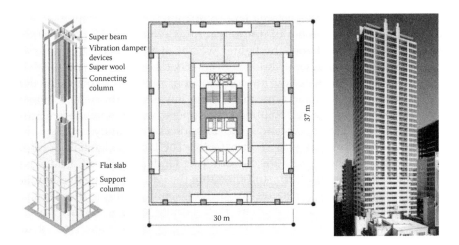

FIGURE 6.37
Super reinforced concrete frame construction.

rebuilt after fifty years, before their physical lives had been fulfilled, and while they were otherwise still sound. This has not only caused enormous waste, but also demanded new resources and energy necessary for renewal activities, thereby contributing to worsening global environmental issues.

The recent development of ultra-high strength concrete has made it possible to decrease the profiles of reinforced concrete (RC) members and realize long-span structures, thereby enabling development of super high rise RC residential buildings with the support and infill separated.

This solved the space issues mentioned earlier and extended building service life. Figure 6.37 shows an example of such super RC frame construction methods used to create a flat open plan by eliminating columns and beams from the living space and supporting the entire building with a core consisting of so-called super walls and super beams at the top. The super beams are connected to peripheral columns through seismic vibration damper devices. This method, as shown in Figure 6.38, allows the free alteration of floor plans as required along with changes in occupancy, while making it possible to provide a different floor-to-ceiling height for each story.

Super RC frame buildings therefore represent typical examples of adaptable buildings in seismic regions, thereby contributing to a sound material-cycle society. Figure 6.39 shows the results of a comparison based on trial calculations of natural resource consumption and waste generation for a conventional RC residential building and super RC frame residential building, assuming that the former had a life of fifty years with one reconstruction, while the latter had a life of one hundred years. This reveals that both natural resource consumption and waste generation can be significantly reduced in the case of the super RC building.

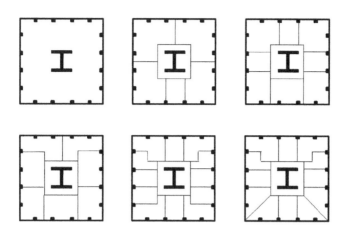

FIGURE 6.38
Free alteration of floor plans.

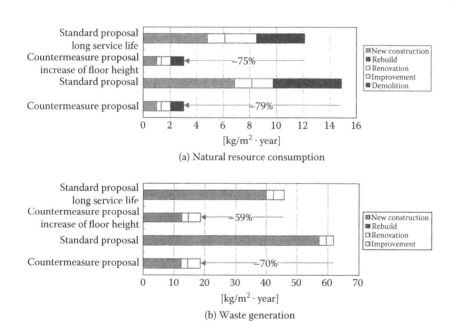

(a) Natural resource consumption

(b) Waste generation

FIGURE 6.39
Natural resource consumption and waste generation of conventional and super reinforced concrete frame buildings.

6.8 Utilization of Thermal Mass of Concrete

In some cases, concrete cannot be necessarily considered an eco-friendly material because its key material (cement) produces immense amounts of CO_2 during its production and also because it uses large numbers of rocks of relatively high density as an aggregate that generates considerable CO_2 emissions from combustion of fossil fuel during transportation. In addition, CO_2 emissions resulting from the consumption of energy for heating, cooling, and lighting a concrete building account for approximately 20% of the total.

In this regard, zero energy buildings are currently attracting attention and are intended to reduce CO_2 emissions from buildings in use to close to zero through the utilization of natural energy. As the effective use of high thermal capacity concrete enhances its feasibility, concrete can play a role in the realization of comfortable and pleasant interior spaces. An example of a zero energy building is the 1,600 m^2 two-story Carlton House studio and reception area of the Gifford headquarters completed in 2004 (Figure 6.40).

The studio has a ground floor made up mostly of an in situ concrete slab while the first-floor consists of a hybrid slab, i.e., precast concrete planks with a post-tensioned in situ topping. The exposed thermal mass of the concrete

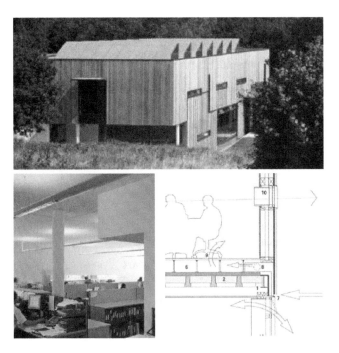

FIGURE 6.40
Carlton House, United Kingdom.

TABLE 6.14

Reduced Energy Consumption and CO_2 Emissions

	Electric Energy (kWh/m²)	Gas Energy (kWh/m²)	Total Energy (kWh/m²)	Total CO_2 (kg/m²)
Typical	147	168	315	95
Best practice	89	95	184	56
Gifford Studios	34	29	63	20

slab is an integral part of the building services design. The concrete floor and exposed concrete soffit allow heating and cooling of the thermal mass. This is achieved by operating the underfloor ventilation system.

Fresh air is supplied by low pressure fans and is cooled or warmed by passing over an underfloor heating and cooling system integrated within the raised access floor. During the daytime, the cooled air stored in the floor slab tempers the incoming fresh air. In the winter, the fans are speed controlled by an air-quality (carbon dioxide) sensor that helps deliver the necessary amount of air with minimal heating load.

The raised floor also houses a piped cooling system to supplement the passive effects of the thermal mass; brackets fixed to the top of the floor pedestals support pre-formed insulation channels carrying plastic pipe circuits fed from zone manifolds. Floor tiles rest on the pedestals and are in direct contact with the pipes. In the summer, water is circulated at 15°C through the underfloor system delivering 30 to 40 W/m² of cooling.

The same system is supplied with warm water at 50°C in the winter to provide heating. The system is served by an air-to-water heat pump that can operate in a heating or cooling mode, depending on the requirements of the space. In addition, thermal mass is enhanced through the use of appropriate lining boards at the roof level and a highly insulated robust breathing system is applied for the wall. Through these innovations, the new studio became one of the most energy efficient buildings in the UK, as shown in Table 6.14.

6.9 Pervious Concrete

The heat island phenomenon was not such an issue when most of the ground surface was covered with particulate materials such as earth and sand, thanks to their adiabatic and evaporation properties. However, with the advancement of urbanization, the once natural ground surface has been covered with paving materials typically fabricated from concrete and asphalt. As a result, heat has been accumulated in large heat storage capacity paving materials that have lower evaporation cooling effects. This is how the heat island phenomenon emerged.

Among the technologies that mitigate the heat island phenomenon through paving materials, one aims to reduce the heat absorption of paving materials by making the color of the pavement surface lighter and enhancing solar reflectance. The solar reflectance index (SRI) is used to assess the mitigation effect, and an eco-friendly building rating system, the Leadership in Energy and Environmental Design (LEED), recognizes that only materials with high reflectances of SRI 29 or more contribute to mitigation of the heat island phenomenon, while the effect of pervious concrete is presently not appreciated. Although the SRI value of pervious concrete is 14 while that of conventional concrete is 37, pervious concrete has properties other than solar reflectance, including low heat capacity, and high adiabatic and evaporation cooling properties, thereby contributing to amelioration of the heat island phenomenon.

A comparison was made between the pervious concrete pavement (consisting of layers of a 15-cm thick pervious concrete with void ratio of 31% and a 45% thick compacted limestone) and conventional concrete pavement (consisting of layers of a 15-cm thick ordinary concrete and compacted soil sub-base) in a parking area of the Iowa State University laid during the summer and autumn of 2006 (Figure 6.41). A comparison was carried out for seven consecutive days with no rain and a temperature of more than 32°C.

Figure 6.42 shows the changes in air temperature of a typical day and the temperature at the medium depth of the roadbed. As the heat capacity, solar reflectance, and thermal conductivity of pervious concrete are smaller than those of conventional concrete, the daytime temperature of the surface of its pavement was 5° higher than the conventional concrete pavement during the intensely hot hours, but there was no significant difference in their nighttime temperature.

FIGURE 6.41
Iowa State University parking area paved with pervious concrete.

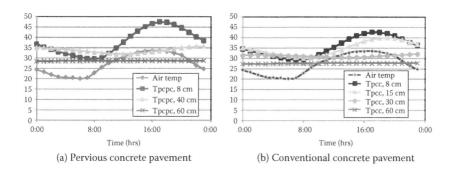

(a) Pervious concrete pavement   (b) Conventional concrete pavement

FIGURE 6.42
Air temperature fluctuations.

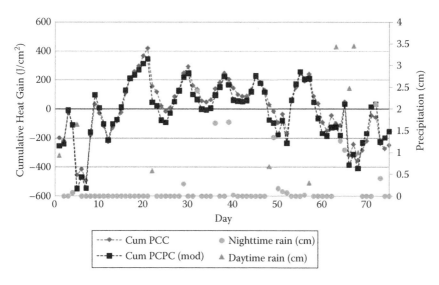

FIGURE 6.43
Heat gain peak.

As shown in Figure 6.43, because the duration of heat gain peaks of the pervious concrete pavement is shorter compared with conventional concrete, and big heat loss follows such peaks, pervious concrete cools more rapidly. This tends to become prominent especially when it rains during the daytime because of the evaporation effect. With its large number of voids, pervious concrete has less cumulative heat storage than conventional concrete as shown in Table 6.15, which demonstrates its potential contribution to mitigation of the heat island phenomenon.

TABLE 6.15

Heat Storage of Pervious Concrete and Conventional Concrete Pavement

Date	Energy Stored in Conventional Concrete (J/cm^2)	Energy Stored in Pervious Concrete (J/cm^2)
2007-7-7	560.2	492.3
2007-7-8	576.7	516.8
2007-7-17	501.6	449.5
2007-7-18	398.4	352.4
2009-8-11	486.6	449.3

References

Haselbach, L., Boyer, M., Kevern, J.T., and Schaefer, V.R. 2011. Cyclic Heat Island Impacts on Traditional versus Pervious Concrete Pavement Systems. Transportation Research Record: Journal of the Transportation Research Board. No.2240. pp.107–115.

Japan Cement Association. 2009. Report on the life cycle cost of concrete pavement. Report R-24 of Committee on Pavement.

JSCE. 2006. Recommendation of environmental performance verification for concrete structures. Guidelines for Concrete No. 7. Japan Society of Civil Engineers.

Kevern, J.T., Haselbach, L., and Schaefer, V.R. 2009. Hot Weather Comparative Heat Balances in Pervious Concrete and Impervious Concrete Pavement Systems. Second International Conference on Countermeasures to Urban Heat Islands. Berkeley.

Kuroda, Y., Hashida, H., and Tsuchiya, M. 2002. Application of recycled aggregate concrete to a structure. *Proceedings of Structural Engineers World Congress*, Yokohama, pp. 1–8.

Sakai, K. 2009. Towards environmental revolution in concrete technologies. *Proceedings of 11th Annual fib Symposium, Concrete: 21st Century Superhero*, London, pp. 13–41.

Taylor, G.W. and Patten, J.D. 2006. Effects of pavement structure on vehicle fuel consumption. Phase III test report. National Research Council of Canada, Technical Report CSTT-HVC-TR-068.

7

Future Perspectives

7.1 Existing Perspectives

7.1.1 General

It is extremely difficult to predict the future. However, unless we make efforts to establish a signpost by which to foresee the future, we will be lost and unable to grasp which direction to take. Hardly any attempts have been made to establish such signposts with respect to concrete. In Chapter 7, we examine a roadmap concerning concrete sustainability and introduce the IEA Energy Technology Perspectives (2010a) and Cement Technology Roadmap (IEA-WBCSD, 2009).

7.1.2 IEA Energy Technology Perspectives

The IEA (2010a) proposed a CO_2 reduction scenario and strategy up to the year 2050 as Energy Technology Perspectives (ETPs) that focus on how low carbon energy technologies can contribute to deep CO_2 emission reduction targets. Today, CO_2 emissions from fossil fuel combustion are estimated as approximately 28 billion tonnes, and the IEA projects that they will reach 57 billion tonnes by 2050 if no countermeasures are taken. As a scenario to halve the present emission level to 14 billion tonnes, low carbon technologies together with their ratios are introduced as follows:

1. Carbon capture and storage (CCS) 19%
2. Renewables 17%
3. Nuclear 6%
4. Power generation efficiency and fuel switching 5%
5. End-use fuel switching 15%
6. End-use fuel and electricity efficiency 38%

These were calculated based on the minimum cost pathways, while applying the techno-economic approach that assesses costs and benefits. The greatest expectation is placed on end-use fuel and electricity efficiency. CCS is positioned as the most important technology to directly reduce CO_2 emissions in the industries including steel, cement, chemical and petrochemical, and pulp and paper. ETP states that "Many of the most promising low-carbon technologies currently have higher costs than the fossil-fuel incumbents. It is only through technology learning from research, development, demonstration and deployment (RDD&D) that these costs can be reduced and the technologies become economic." The renewable technologies would be top of the list.

CO_2 emissions from buildings account for about 10% of the global total. ETP demonstrates that as most buildings have long life spans, more than half of the current global building stock will still be used in 2050, and "while the CO_2 savings potential of developed nations lies in retrofitting and purchasing new technologies for the existing building stock, in developing countries where building growth will be very rapid, opportunities exist to secure significant energy savings through improved efficiency standards for new buildings." In conclusion, "the most vital message of ETP 2010 is that an energy technology revolution is within reach."

7.1.3 IEA-WBCSD Cement Technology Roadmap

In 2009, the Cement Sustainability Initiative (CSI) of the World Business Council for Sustainable Development (WBCSD) and IEA jointly presented a roadmap regarding the contribution by the cement industry to halve CO_2 emissions by 2050. This shows that the cement industry has the following four carbon emissions reduction levers:

1. Thermal and electric efficiency including deployment of existing state-of-the-art technologies in new cement plants and retrofits of energy efficiency equipment where economically viable
2. Alternative fuels, i.e., use of less carbon-intensive fossil fuels and more alternative fuels (wastes) and biomass fuels
3. Clinker substitution, i.e., materials with cementitious properties
4. CCS

The roadmap demonstrates that energy intensity, alternative fuel use, and cement-to-clinker ratios should have changed in 2010 from 3.9 to 3.2 GJ/t clinker, 5% to 37% and 77% to 71%, respectively. With respect to CCS technology, it is envisaged that it will be introduced in 220 to 430 cement kilns. In addition, regarding 2.34 billion tonnes of CO_2 emissions projected in the case that no special measures are taken, emissions can be reduced to 1.55 billion tonnes, lower by about one-third, by combining the four measures. The reduction ratios of the levers are as follows:

Energy efficiency 10%

Alternative fuel use and other fuel switching 24%

Clinker substitution 10%

CCS 56%

These ratios reveal that it is difficult to drastically reduce CO_2 emissions by cement manufacturing technologies alone and that there is little choice but to rely on CCS technology for more than half.

7.2 New Perspectives

7.2.1 General

The Brundtland report was published in 1987 by the United Nations World Commission on Environment and Development, and the report defined sustainable development for the first time. It stated that all nations must set goals of economic and social development based on sustainability. The concept of sustainability represents the most important discovery of the twentieth century, which can be called the third revolution to fundamentally review the fruits of the Agricultural Revolution and Industrial Revolution, from which mankind has benefited. The essence of sustainability is to meet the needs of both the present generation and also future generations. In order to satisfy this, it is important to sustain the global environment without causing depletion of its limited resources and energy.

It is easily understandable that the exploitation of natural resources and energy normally conflicts with sustainment of the environment. Nevertheless, over 250 years following the Industrial Revolution, mankind has fortunately come to realize that such conflicting events have to be made compatible under the concept of sustainability. However, standing firmly in the way of realizing sustainability are (1) the existing system with an economic development model based on expansion of resources and energy exploitation and (2) the economic activities of developing countries that will surely show significant economic growth in the future. Furthermore, disasters caused by climate change and earthquakes due to tectonic shifts inflicted serious damage in some regions, frequently causing suspension of social and economic activities.

Sustainability is described as a concept that includes environmental, economic and social aspects. As these aspects interrelate, the concept of sustainability is in reality extremely complex and is not likely to have been clearly explained. This is also due to the fact that until today, in most cases, sustainability has been associated with the environmental aspect. For example, reduction of CO_2 emissions, regarded as the root cause of global warming, has been the top priority concerning sustainability of the Earth. This is because global climate is the most important factor concerning mankind's survival on the Earth, a celestial sphere whose radius is only 6,300 km. The Earth including its atmosphere comprises an extremely complex system, and it is not difficult to imagine the enormous danger attached to changing the rhythm of nature through the excessive activities of mankind.

Therefore environmental sustainability is the foundation for mankind's survival. Although we carry out various activities within such restrictions related to environmental sustainability, we at times forget these important restrictions or ignore them for our own immediate benefit. However, the occasional occurrence of catastrophe reminds us of what kind of environment we live under, although as time passes, we again forget or ignore it. While we have repeated this, we have also taken various countermeasures by learning from disasters. Now mankind also faces challenges concerning the depletion of resources and energy.

Thus there is no right answer to the proposition of sustainability; it is like trying to solve a multiple simultaneous equation with loaded influential factors. What we can do is to seek the best solution by examining various assumptions. Now, the authors would like to clarify current problems and review the issue of sustainability.

7.2.2 Future Direction of Sustainability

The Great East Japan Earthquake on March 11, 2011, presented us with an opportunity to think about the essence of the issue of sustainability. We have been making use of resources and energy to build a social and economic infrastructure on which we base production and cultural and social activities. However, the major earthquake and the huge tsunami it triggered destroyed much of this in an instant. On top of it came radioactive contamination caused by a serious accident at one of our nuclear power stations, the energy source which has been presented as the key player in mitigating global warming.

As the Japanese archipelago lies on the joining of multiple tectonic plates, earthquakes are unavoidable parts of its destiny. In the construction field, safer structural design has been implemented whenever damage has been caused by a large earthquake. The March 2011 earthquake once again demonstrated to people the importance of infrastructural safety and reliability including in construction and housing. In other words, tragic experience has once more taught us that, without a robust infrastructure, human society has no chance of sustainability.

Huge amounts of resources and energy are used for infrastructure development, requiring massive investment. As such development inevitably requires land use and land modification, it brings with it natural destruction.

Sustainability needs to be evaluated from the environmental, economic, and social perspectives. The March 2011 disaster showed very graphically its destructive effects in each of these areas. The natural movements of the crustal deformation produced extensive changes in the natural environment. In other words, environments formed by nature were destroyed by nature, and as a result, environments that humans made by modifying nature's surface were destroyed by the tsunami and liquefaction.

Man-made objects suffered earthquake damage of varying degrees. The tsunami claimed approximately 20,000 victims. Moreover, at a supposedly robustly constructed nuclear power station, the loss of electric power led to interruption of the supply of cooling water so that fuel rods went into meltdown, and the resulting hydrogen gas caused an explosion that damaged the building and allowed radioactive substances to scatter over a wide area. It is feared that this radioactive contamination will have damaging effects on human health, and the prediction is that it will ultimately take ten to twenty years to deal with the consequences of the accident.

Nuclear power has been promoted as being free from CO_2 emissions and representing a low-cost form of electricity generation, but this accident undermined the basis of that argument. Nuclear power generation, which is at the heart of our response to the energy problem, is beginning to be viewed very critically. In addition to the environmental issue of radioactive contamination, this accident revealed grave social and economic problems, notably the long-term evacuation of the local population and the suspension of local economic activity. The main cause of this situation, which ranks as an environmental crisis, is that the nuclear power station facility was not designed to withstand a major tsunami. In other words, the question concerns facility design.

The people of the regions hit by the earthquake and tsunami lost their homes, and the region's production activities suffered very serious damage. Some regions had clusters of plants producing automotive and electronic parts that served the world. Although there are many enterprises that have managed speedy recoveries, it seems likely that the disaster will accelerate the overseas transfer of manufacturing.

In addition to loss of human life, buildings, and infrastructure, disasters thus greatly impair social and economic activities. The greater the scale of the disaster and the wider its extent, the greater the impact. The March 2011 disaster produced many victims, dealt the economy a severe blow, brought societal activity to a complete standstill, and created new environmental issues by necessitating massive resource and energy deployments to clear large volumes of rubble and rebuild destroyed buildings and infrastructure that are necessary for recovery. There is no way of calculating the economic loss, but it is thought likely that it will ultimately be equivalent to the Japanese state budget for a whole year.

In this way, the destruction of social infrastructure and facilities exerts a lethal impact on economic and social activities. Social infrastructure and facilities serve as bases for social and economic activities. Their availability enhances social and economic activities and, according to basic economic principles, encourages expansion of all activities. Thus simple expansion

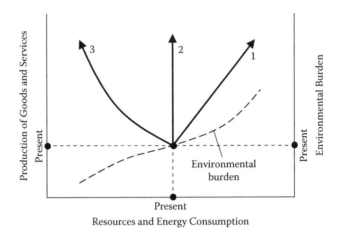

FIGURE 7.1
Three environmental burden scenarios.

of economic activity is not desirable from a sustainability perspective. Nevertheless, effecting a major increase in resource and energy efficiency to facilitate economic expansion is the ultimate goal. Figure 7.1 is a diagrammatic representation of the consumption of resources and energy, the production of goods and services, and their relation to environmental burden based on three different scenarios.

Scenario 1: Resource and energy efficiency stays at current levels, while population increases and living standards rise.

Scenario 2: Population increases and living standards rise, and resource and energy efficiency rises in proportion to them.

Scenario 3: Population increases and living standards rise, while the increase in resource and energy efficiency is greater than the increase of population and living standards.

In Scenario 1, the environmental burden increases while in Scenario 2 it is maintained at the current level. However, it is recognized that Scenario 2 is an inadequate solution under current global conditions. We therefore need to think of a future direction based on Scenario 3, in which resource and energy efficiency is drastically improved, the environmental burden is significantly lowered, and the volume of goods and services produced is also increased above the present level.

Within the sustainability framework outlined above, the position of the concrete and construction fields needs to be clarified. In the construction field, whenever an earthquake or other major disaster has caused damage, structural design has been made safer. The March 2011 disaster again caused people to recognize the importance of safety and reliability in buildings and infrastructure. In other words, they have learned that robust buildings and infrastructure are the foundations of sustainability.

Scenario 3 above refers not only to general production activity but also includes the resources and energy consumed in the concrete and construction fields. If the population increases and living standards rise, demand for buildings and infrastructure development increases, and as a direct or indirect consequence, resource and energy consumption grows. It would be premature to think that the March 2011 events have swept away the problem of reducing environmental burden; the need to consider resources and energy (the environmental aspect) and the economy as well as safety and reliability (the social aspect) as part of the sustainability framework, remains.

Striking a balance between these two issues is far from simple. In fact, there are major issues regarding the robustness of existing buildings and infrastructure, and when it comes to the minimization of resource and energy use in future buildings and infrastructure development, the available technology and assessment rules are a long way from adequate. What we can do is learn from experience and keep our eyes on the future as we continue working to resolve the issues one at a time.

7.2.3 Conflicts of Structural Safety-Reliability, Environmental Issues, and Future Direction

In design for buildings and infrastructures, assumptions of load and environmental actions are generally made to verify whether adequate safety can be ensured during the intended service life of a structure. The structural design method has evolved successively from the allowable stress design method to the limited state design method and now the performance-based design method. This resulted from improvement in the accuracy with which we can determine the quality of the materials used and the durability and load resistance behavior of structures.

The basic concept of performance-based design, which is nowadays becoming common, is that as long as the required performance is attained, any materials or structural format is acceptable. This is based on the assumption that the behavior of a structure can be followed with a fair degree of accuracy. In performance-based design, the question is how much redundancy (safety) to incorporate into assumed actions, depending on which the behavior of the structure differs completely.

If a large degree of redundancy is adopted, the degree of safety increases in respect of actions exceeding the assumed level. However, this generally increases the amounts of resources and energy used. On the other hand, if a low level of redundancy is adopted, highly efficient resource and energy consumption results, but the degree of safety decreases in respect of actions exceeding the assumed level. This is a conflict between structural safety and reliability on one hand and environmental issues on the other. So far, this problem has been very little discussed.

The fact is that it is impossible to create a design that completely excludes destruction by great external forces that have not been experienced previously but may occur in the future; it is also not economically practicable. It may therefore sometimes be necessary to provide a level of performance which, in the case of an unforeseen major action of very low probability, holds damage to the level of partial destruction without collapse of the entire structure. Such level of performance is called robustness. Although much wooden housing was destroyed by the tsunami in March 2011, reinforced concrete buildings escaped destruction, and many people's lives were saved by taking refuge on rooftops.

Robustness is also crucially important in infrastructure. In the March 2011 disaster, the columns and beams of the viaduct that carries the Tohoku Shinkansen bullet train suffered damage, but it amounted only to localized destruction and did not lead to the collapse of the viaduct as a whole. As a result, train accidents were prevented, and the viaduct resumed operation within a short time.

Figure 7.2 is a diagrammatic representation of the relation of degree of safety to environmental burden based on three different scenarios.

   Scenario 1: Increasing the degree of safety under current technology
      leads to an increase in environmental burden.

   Scenario 2: Improvements to materials and structure allow the degree
      of safety to be increased without increasing environmental burden.

   Scenario 3: Introduction of innovative technology in materials and
      structure reduces environmental burden and increases safety.

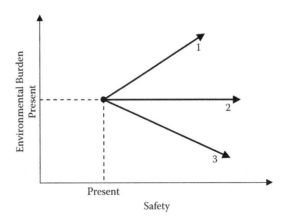

FIGURE 7.2
Three scenarios relating safety to environmental burden.

Scenario 1 is the case in which the degree of safety can be improved by, for example, increasing cross sections and reinforcing bars, which inevitably augments environmental burden. However, this can be an option for very important structures. Scenarios 2 and 3 assume the introduction of technology to reduce environmental burden, but it is essential to additionally integrate the concept of robustness into designs in an appropriate fashion. The future direction toward which we should aim is quite clearly Scenario 3. A range of levels of robustness can be envisaged, but major factors will be protection of human life, facilitating recovery, and the environmental burden caused by demolition and new construction.

7.2.4 Energy Issues

The most serious issues resulting from the March 2011 disaster are the radioactive contamination from the nuclear power station accident and the task of dealing safely with the nuclear power fuel that went into meltdown. According to the IEA (2010b), nuclear power generation in 2008 represented only 13.5% of the world total. Coal, gas, and hydroelectric power accounted, respectively, for 41%, 21.3%, and 15.9%. In world terms, the share of nuclear power electricity generation is thus quite small. The number of nuclear power stations worldwide is more than 430, and the countries with the most are America, France, and Japan with 104, 59 and 54, respectively. In Japan, nuclear power previously accounted for 30% of total electricity generation, but due to the March 2011 disaster and scheduled inspections, only 2 nuclear power stations were in operation as of February 2012.

In the wake of the disaster, the world has come to regard nuclear power generation with a very critical eye, and it is not clear how the future will develop. However, one thing that has become clear once again is that, with nuclear power, once an incident has occurred it becomes extremely hard to control. Particularly in countries such as Japan with a small land area and a dense population, there is a high risk of radioactive contamination having a very great impact.

As Japan is a major energy-consuming country that has almost no coal or petroleum resources, the shift to nuclear power generation was a natural progression. However, uranium, the fuel used in nuclear power generation, is a resource expected to last for only around one hundred years, and the issue of radioactive waste treatment has not been resolved. There we face a daunting range of problems. Coal and petroleum meanwhile, in light of the future energy consumption of China and India, are likely to surge in price and cause global warming to accelerate. However, in view of the way in which energy consumption has expanded up to now, it would not be practical to halt nuclear power generation with immediate effect. Establishing a still higher level of safety in nuclear power generation while progressively expanding investment in renewable energy development is the only way forward.

In the construction field, the use of coal in steel and cement production is a major source of energy consumption. As transportation and construction use mainly petroleum-based fuel, direct consumption of electricity is likely to be small. CO_2 emissions also are to a large extent related to the manufacture of materials. However, given the surge in the price of coal and the need to limit CO_2 emissions, development of innovative technology for steel and cement manufacture is essential. Moreover, electricity consumption needs to be further tackled through energy conservation and introduction of renewable energy.

7.3 Roadmap toward Sustainable Concrete Industry

It is not yet 200 years since modern cement was invented in Britain by Aspdin. However, due to its outstanding properties, its use has expanded rapidly so that nowadays it is one of the most useful construction materials and has become an indispensable resource. Its use is certain to carry on expanding in the future. A vast body of information on concrete technology has accumulated over the years. However, at the end of the twentieth century, mankind discovered the concept of sustainability.

In other words, it was realized that, because of the Earth's limited resource and energy reserves and the risk of climate change, we need to apply a completely new set of values to production activity than values applied up to now. Mankind is therefore at the start of a third revolution to follow the Agricultural Revolution and the Industrial Revolution. Let us call it the Sustainability Revolution. To implement the Sustainability Revolution in the concrete industry, great efforts will be needed; the roadmap will probably look something like this:

1. Ascertain the current situation in the concrete industry regarding consumption of resources and energy and CO_2 emissions, and perform continuous monitoring. (Start immediately and progressively improve the accuracy of the data).

2. Establish a new system by integrating an environmental aspect into current specifications and standards for the design of concrete structures (by 2020).

3. Promote technology development to reduce resource and energy consumption in the concrete industry and in parallel build a system to promote use of the newly developed technology (by 2020).

4. Develop relevant indices effective in environmental burden assessment (by 2015).

5. As a goal for 2020, reduce environmental burden by 20 to 30% compared to 2010 through use of BAT.

6. At the 2020 mark, assess the achievements of the previous ten years and set goals for the next ten years, ideally, a goal of 50 to 60% reduction compared to 2010 (by 2030).

7. It is important for international organizations (fib, ACI, ACF, etc.) that have already started such activities to coordinate. Action with a view to establishing a relevant forum is required.

The roadmap described here is not based on a detailed model and is no more than a very rough orientation. However, we think that the main aims we need to set are clear. We emphasize that in the future, a more concrete assessment will be needed.

7.4 Wishful Thinking

We live in what is in some ways a wonderful age, and enjoy the fruits of the Earth to the extent that it becomes smaller. However, while this unique Earth planet is extremely rich, it is also extremely fragile. This has been brought home to us by the experience of watching Earth from space.

The inexhaustible appetites that mankind has acquired since the Industrial Revolution have produced a wide range of problems. On the other hand, these problems have been acknowledged and efforts to resolve them are being made. In other words, mankind has a history of repeatedly overstepping the mark, learning the lesson, and correcting its course. Put simply, sustainability, the keyword of the twenty-first century, means "not overstepping the mark."

It is absolutely natural that mankind has used the materials at hand to create buildings and infrastructure and thereby build a comfortable and convenient environment. This does not fundamentally change even when society becomes complex and sophisticated. Nevertheless, for the Earth's population of 7 to 10 billion to lead an equally comfortable life, we need massive amounts of resources and energy, and the attitudes and technology of the twentieth century have clearly caused the Earth to suffer degradation.

For our and future generations, the direction we need to take in the twenty-first century is to create conditions that allow us to live in a sustainable fashion and at the same time improve the quality of life. In contemporary society, almost all people utilize infrastructure and spend long periods of time inside houses and buildings. Creating a human-friendly and safe environment with low environmental burden is the responsibility of those of us working in the concrete and construction industry. Let us continue to work boldly and continuously toward that noble goal.

References

IEA. 2010a. *Energy Technology Perspectives: Scenarios & Strategies to 2050*. International Energy Agency, Paris.

IEA. 2010b. *Key World Energy Statistics*. International Energy Agency, Paris.

IEA-WBCSD. 2009. *Cement Technology Roadmap 2009: Carbon Emissions Reduction up to 2050*. International Energy Agency, Paris.

Index

Printed and bound by CPI Group (UK) Ltd, Croydon, CR0 4YY

18/10/2024

01776269-0011